PIPELINE DESIGN FOR WATER ENGINEERS

THIRD REVISED AND UPDATED EDITION

PIPELINE DESIGN FOR WATER ENGINEERS

THIRD REVISED AND UPDATED EDITION

DAVID STEPHENSON

Department of Civil Engineering
University of the Witwatersrand
Johannesburg, South Africa

ELSEVIER

Amsterdam — Oxford — New York — Tokyo 1989

ELSEVIER SCIENCE PUBLISHERS B.V.
Sara Burgerhartstraat 25
P.O. Box 211, 1000 AE Amsterdam, The Netherlands

Distributors for the United States and Canada:

ELSEVIER SCIENCE PUBLISHING COMPANY INC.
655, Avenue of the Americas
New York, NY 10017, U.S.A.

ISBN 0-444-87373-2

Printed in The Netherlands

PREFACE TO FIRST EDITION

Pipelines are being constructed in ever-increasing diameters, lengths and working pressures. Accurate and rational design bases are essential to achieve economic and safe designs. Engineers have for years resorted to semi-empirical design formulae. Much work has recently been done in an effort to rationalize the design of pipelines.

This book collates published material on rational design methods as well as presenting some new techniques and data. Although retaining conventional approaches in many instances, the aim of the book is to bring the most modern design techniques to the civil or hydraulic engineer. It is suitable as an introduction to the subject but also contains data on the most advanced techniques in the field. Because of the sound theoretical background the book will also be useful to under-graduate and post-graduate students. Many of the subjects, such as mathematical optimization, are still in their infancy and the book may provide leads for further research. The methods of solution proposed for many problems bear in mind the modern acceptance of computers and calculators and many of the graphs in the book were prepared with the assistance of computers.

The first half of this book is concerned with hydraulics and planning of pipelines. In the second half, structural design and ancillary features are discussed. The book does not deal in detail with manufacture, laying and operation, nor should it replace design codes of practice from the engineer's desk. Emphasis is on the design of large pipelines as opposed to industrial and domestic piping which are covered in other publications. Although directed at the water engineer, this book will be of use to engineers involved in the piping of many other fluids as well as solids and gases.

It should be noted that some of the designs and techniques described may be covered by patents. These include types of pre-stressed concrete pipes, methods of stiffening pipes and branches and various coatings.

The S.I. system of metric units is preferred in the book although imperial units are given in brackets in many instances. Most graphs and equations are represented in universal dimensionless form. Worked examples are given for many problems and the reader is advised to work through these as they often elaborate on ideas not highlighted in the text. The algebraic symbols used in each chapter are summarized at the end of that chapter together with specific and general references arranged in the order of the subject matter in the chapter. The appendix gives further references and standards and other useful data.

PREFACE TO SECOND EDITION

The gratifying response to the first edition of this book resulted in small amendments to the second impression, and some major alterations in this new edition.

The chapters on transport of solids and sewers have been replaced by data more relevant to water engineers. Thus a new chapter on the effects of air in water pipes is included, as well as a chapter on pumping systems for water pipelines. The latter was reviewed by Bill Glass who added many of his own ideas.

There are additions and updating throughout. There is additional information on pipeline economics and optimum diameters in Chapter 1. A comparison of currently used friction formulae is now made in Chapter 2. The sections on non-circular pipe and partly full pipes and sewer flow are omitted. These are largely of interest to the drainage engineer and as such are covered in the author's book 'Stormwater Hydrology and Drainage' (Elsevier, 1981). A basic introduction to water hammer theory preceeds the design of water hammer protection of pumping and gravity lines in Chapter 4.

The sections on structural design of flexible pipes are brought together. An enlarged section on soil-pipe interaction and limit states of flexible pipes preceeds the design of stiffened pipes.

Although some of the new edition is now fairly basic, it is recognised that this is desirable for both the practicing engineer who needs refreshing and the student who comes across the problem of pipeline design for the first time.

PREFACE TO THIRD EDITION

Recent research in cavitation and flow control has prompted additional sections on this. There are also new sections on supports to exposed pipes and secondary stress. Additional references and a new layout make up this edition. Some sections appearing in previous editions, noteably on pipe network systems analysis and optimization have been ommitted as they were considered more appropriate in the author's parallel book 'Pipeflow Analysis' by the same publisher.

ACKNOWLEDGEMENTS

The basis for this book was derived from my experience and in the course of my duties with the Rand Water Board and Stewart, Sviridov and Oliver, Consulting Engineers. The extensive knowledge of Engineers in these organizations may therefore be reflected herein although I am solely to blame for any inaccuracies or misconceptions.

I am grateful to my wife Lesley, who, in addition to looking after the twins during many a lost weekend, assiduously typed the first draft of this book.

David Stephenson

X

CONTENTS

CHAPTER 1   ECONOMIC PLANNING

CHAPTER 2   HYDRAULICS

CHAPTER 3   PIPELINE SYSTEM ANALYSIS AND DESIGN

CHAPTER 7   CONCRETE PIPES

CHAPTER 8   STEEL AND FLEXIBLE PIPE

CHAPTER 9   SECONDARY STRESSES

CHAPTER 10   PIPES, FITTINGS AND APPURTENANCES

CHAPTER 11   LAYING AND PROTECTION

CHAPTER 12   PUMPING INSTALLATIONS

CHAPTER 1

ECONOMIC PLANNING

INTRODUCTION

Pipes have been used for many centuries for transporting fluids. The Chinese first used bamboo pipes thousands of years ago, and lead pipes were unearthed at Pompeii. In later centuries wood-stave pipes were used in England. It was only with the advent of cast iron, however, that pressure pipelines were manufactured. Cast iron was used extensively in the 19th Century and is still used. Steel pipes were first introduced towards the end of the last century, facilitating construction of small and large bore pipelines. The increasing use of high grade steels and large rolling mills has enabled pipelines with diameters over 3 metres and working pressure over 10 Newtons per square millimetre to be manufactured. Welding techniques have been perfected enabling longitudinally and circumferentially welded or spiral welded pipes to be manufactured. Pipelines are now also made in reinforced concrete, pre-stressed concrete, asbestos cement, plastics and claywares, to suit varying conditions. Reliable flow formulae became available for the design of pipelines this century, thereby also promoting the use of pipes.

Prior to this century water and sewage were practically the only fluids transported by pipeline. Nowadays pipelines are the most common means for transporting gases and oils over long distances. Liquid chemicals and solids in slurry form or in containers are also being pumped through pipelines on ever increasing scales. There are now over two million kilometres of pipelines in service throughout the world. The global expenditure on pipelines in 1974 was probably over £5 000 million.

There are many advantages of pipeline transport compared with other methods such as road, rail, waterway and air:-

(1) Pipelines are often the most economic form of transport (considering either capital costs, running costs or overall costs).

(2) Pipelining costs are not very susceptible to fluctuations in

prices, since the major cost is the capital outlay and subsequent operating costs are relatively small.

(3) Operations are not susceptible to labour disputes as little attendance is required. Many modern systems operate automatically.

(4) Being hidden beneath the ground a pipeline will not mar the natural environment.

(5) A buried pipeline is reasonably secure against sabotage.

(6) A pipeline is independent of external influences such as traffic congestion and the weather.

(7) There is normally no problem of returning empty containers to the source.

(8) It is relatively easy to increase the capacity of a pipeline by installing a booster pump.

(9) A buried pipeline will not disturb surface traffic and services.

(10) Wayleaves for pipelines are usually easier to obtain than for roads and railways.

(11) The accident rate per ton – km is considerably lower than for other forms of transport.

(12) A pipeline can cross rugged terrain difficult for vehicles to cross.

There are of course disadvantages associated with pipeline systems:-

(1) The initial capital expenditure is often large, so if there is any uncertainty in the demand some degree of speculation may be necessary.

(2) There is often a high cost involved in filling a pipeline (especially long fuel lines).

(3) Pipelines cannot be used for more than one material at a time (although there are multi-product pipelines operating on batch bases).

(4) There are operating problems associated with the pumping of solids, such as blockages on stoppage.

(5) It is often difficult to locate leaks or blockages.

PIPELINE ECONOMICS

The main cost of a pipeline system is usually that of the pipeline itself. The pipeline cost is in fact practically the only cost for gravity systems but as the adverse head increases so the power and pumping station costs increase.

Table 1.1 indicates some relative costs for typical installed pipelines.

With the economic instability and rates of inflation prevailing at the time of writing pipeline costs may increase by 20% or more per year, and relative costs for different materials will vary. In particular the cost of petro-chemical materials such as PVC may increase faster than those of concrete for instance, so these figures should be inspected with caution.

TABLE 1.1   Relative Pipeline Costs

Pipe Material	Bore mm		
	150	450	1 500
PVC	6	23	–
Asbestos cement	7	23	–
Reinforced concrete	–	23	80
Prestressed concrete	–	33	90 – 150
Mild steel	10	28	100 – 180
High tensile steel	11	25	90 – 120
Cast iron	25	75	–

*"–" indicates not readily available.

1 unit = £/metre in 1974 under average conditions

The components making up the cost of a pipeline vary widely from situation to situation but for water pipelines in open country and typical conditions are as follows:-

Supply of pipe	– 55% (may reduce as new materials are developed)
Excavation	– 20% (depends on terrain, may reduce as mechanical excavation techniques improve)
Laying and jointing	– 5% (may increase with labour costs)
Fittings and specials	– 5%
Coating and wrapping	– 2%
Structures (valve chambers, anchors)	– 2%

4

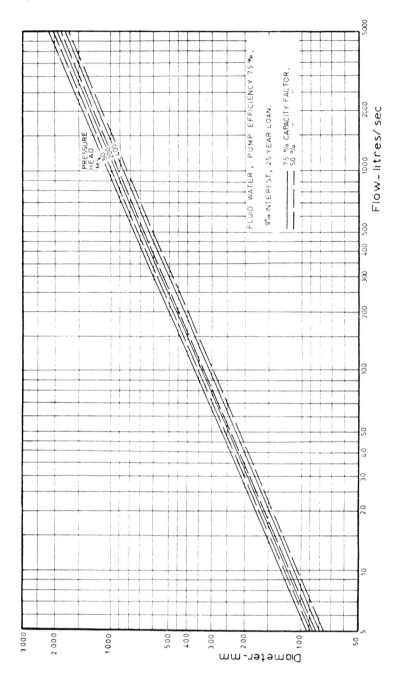

Fig. 1.1 Optimum pumping main diameters for a particular set of conditions

Water hammer protection   – 1%

Land acquisition, access roads, cathodic

protection, security structures, fences   – 1%

Engineering and survey costs   – 5%

Administrative costs   – 1%

Interest during construction   – 3%

Many factors have to be considered in sizing a pipeline: For water pumping mains the flow velocity at the optimum diameter varies from 0.7 m/s to 2 m/s, depending on flow and working pressure. It is about 1 m/s for low pressure heads and a flow of 100 ℓ/s increasing to 2 m/s for a flow of 1 000 ℓ/s and pressure heads at about 400 m of water, and may be even higher for higher pressures. The capacity factor and power cost structures also influence the optimum flow velocity or conversely the diameter for any particular flow. Fig. 1.1 illustrates the optimum diameter of water mains for typical conditions.

In planning a pipeline system it should be borne in mind that the scale of operation of a pipeline has considerable effect on the unit costs. By doubling the diameter of the pipe, other factors such as head remaining constant, the capacity increases six-fold. On the other hand the cost approximately doubles so that the cost per unit delivered decreases to 1/3 of the original. It is this scale effect which justifies multi-product lines. Whether it is in fact economical to install a large diameter main at the outset depends on the following factors as well as scale:-

(1)   Rate of growth in demand (it may be uneconomical to operate at low capacity factors during initial years). (Capacity factor is the ratio of actual average discharge to design capacity).

(2)   Operating factor (the ratio of average throughput at any time to maximum throughput during the same period), which will depend on the rate of draw-off and can be improved by installing storage at the consumer's end.

(3)   Reduced power costs due to low friction losses while the pipeline is not operating at full capacity.

(4)   Certainty of future demands.

(5)   Varying costs with time (both capital and operating).

(6)   Rates of interest and capital availability.

(7)   Physical difficulties in the construction of a second pipeline if required.

The optimum design period of a pipeline depends on a number of factors, not least being the rate of interest on capital loans and the rate of cost inflation, in addition to the rate of growth, scale and certainty of future demands (Osborne and James, 1973).

In waterworks practice it has been found economic to size pipelines for demands up to 10 to 30 years hence. For large throughput and high growth rates, technical capabilities may limit the size of the pipeline, so that supplementation may be required within 10 years. Longer planning stages are normally justified for small bores and low pressures.

It may not always be economic to lay a uniform bore pipeline. Where pressures are high it is economic to reduce the diameter and consequently the wall thickness.

In planning a trunk main with progressive decrease in diameter there may be a number of possible combinations of diameters. Alternative layouts should be compared before deciding on the most economic. Systems analysis techniques such as linear programming and dynamic programming are ideally suited for such studies.

Booster pump stations may be installed along lines instead of pumping to a high pressure head at the input end and maintaining a high pressure along the entire line. By providing for intermediate booster pumps at the design stage instead of pumping to a high head at the input end, the pressure heads and consequently the pipe wall thicknesses may be minimized. There may be a saving in overall cost, even though additional pumping stations are required. The booster stations may not be required for some time.

The capacity of the pipeline may often be increased by installing booster pumps at a later stage although it should be realised that this is not always economic. The friction losses along a pipeline increase approximately with the square of the flow, consequently power losses increase considerably for higher flows.

The diameter of a pumping main to convey a known discharge can be selected by an economic comparison of alternative sizes. The pipeline cost increases with increasing diameter, whereas power cost

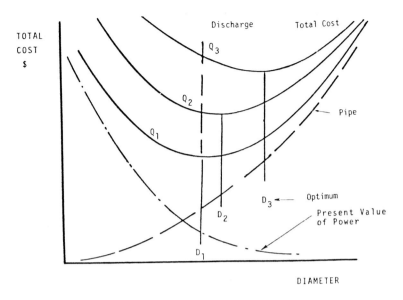

Fig. 1.2   Optimization of diameter of a pumping pipeline.

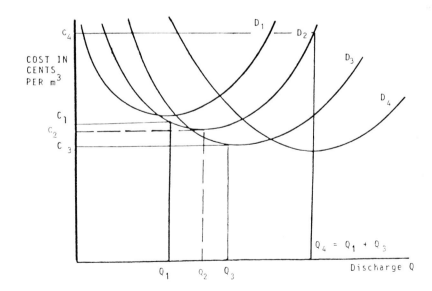

Fig. 1.3   Optimization of throughput for certain diameters

in overcoming friction reduces correspondingly. On the other hand power costs increase steeply as the pipeline is reduced in diameter. Thus by adding together pipeline and the present value of operating costs, one obtains a curve such as Figure 1.2, from which the least-cost system can be selected. An example is given later in the chapter. There will be a higher cost the greater the design discharge rate.

If at some stage later it is desired to increase the throughput capacity of a pumping system, it is convenient to replot data from a diagram such as Figure 1.2 in the form of Figure 1.3. Thus for different possible throughputs, the cost, now expressed in cents per kilolitre or similar, is plotted as the ordinate with alternative (real) pipe diameter a parameter.

It can be demonstrated that the cost per unit of throughput for any pipeline is a minimum when the pipeline cost (expressed on an annual basis) is twice the annual cost of the power in overcoming friction.

Thus the cost in cents per cubic metre of water is

$$C = \frac{C_1(P) + C_2(d)}{Q}$$

$$= \frac{C_1 \, wHQ + C_2(d)}{Q} \tag{1.1}$$

$$= C_1 w \left(\frac{\lambda Q^2 \ell}{2gdA^2} + H_s\right) + C_2(d)/Q$$

$$\therefore \frac{dC}{dQ} = 2C_1 \, wH_f/Q - C_2(d)/Q^2$$

$$= 0 \text{ for minimum C}$$

i.e. $C_2(d) = 2 \, C_1(P)$ \hfill (1.2)

P is power requirement, proportional to wHQ. w is the unit weight of water, H is the total head, subscript s refers to static and f to friction, Q is pumping rate, $C_2(d)$ is the cost of a pipeline of diameter d, $C_1(P)$ is the cost of power (all costs converted to a unit time base).

(In a similar manner it can be shown that the power output of a given diameter penstock supplying a hydroelectric station is a maximum if the friction head loss is one third of the total head available).

Returning to Figure 1.3, the following will be observed:

(1) At any particular throughput Q_1 there is a certain diameter at which overall costs will be a minimum (in this case D_2).

(2) At this diameter the cost per ton of throughput could be reduced further if throughput was increased. Costs would be a minimum at some throughput Q_2. Thus a pipeline's optimum throughput is not the same as the throughput for which it is the optimum diameter.

(3) If Q_1 were increased by an amount Q_3 so that total throughput $Q_4 = Q_1 + Q_3$ it may be economic not to install a second pipeline (with optimum diameter D_3) but to increase the flow through the pipe with diameter D_2, i.e. $Q_4 C_4$ is less than $Q_1 C_1 + Q_3 C_3$.

(4) At a later stage when it is justified to construct a second pipeline the throughput through the overloaded line could be reduced.

The power cost per unit of additional throughput decreases with increasing pipe diameter so the corresponding likelihood of it being most economic to increase throughput through an existing line increases with size (White, 1969).

BASICS OF ECONOMICS

Economics is used as a basis for comparing alternative schemes or designs. Different schemes may have different cash flows necessitating some rational form of comparison. The crux of all methods of economic comparison is the discount rate which may be in the form of the interest rate on loans or redemption funds. National projects may require a discount rate different from the prevailing interest rate, to reflect a time rate of preference, whereas private organizations will be more interested in the actual cash flows, and consequently use the real borrowing interest rate.

The cash flows, i.e. payments and returns, of one scheme may be compared with those of another by bringing them to a common time basis. Thus all cash flows may be discounted to their present value. For instance one pound received next year is the same as £1/1.05 (its present value) this year if it could earn 5% interest if invested this year. It is usual to meet capital expenditure from a loan over a definite period at a certain interest rate. Provision is made for repaying the load by paying into a sinking fund which also collects interest. The annual repayments at the end of each year required to amount to £1 in n years is

$$\frac{r}{(1+r)^n - 1} \tag{1.3}$$

where r is the interest rate on the payments into the sinking fund. If the interest rate on the loan is R, then the total annual payment is

$$\frac{R(1+r)^n + r - R}{(1+r)^n - 1} \tag{1.4}$$

Normally the interest rate on the loan is equal to the interest rate earned by the sinking fund so the annual payment on a loan of £1 is

$$\frac{r(1+r)^n}{(1+r)^n - 1} \tag{1.5}$$

Conversely the present value of a payment of £1 at the end of each year over n years is

$$\frac{(1+r)^n - 1}{r(1+r)^n} \tag{1.6}$$

The present value of a single amount of £1 in n years is

$$\frac{1}{(1+r)^n} \tag{1.7}$$

Interest tables are available for determining the annual payments on loans, and the present values of annual payments or returns, for various interest rates and redemption periods (Instn. of Civil Engs., 1962)

Methods of Analysis

Different engineering schemes required to meet the same objectives may be compared economically in a number of ways. If all payments and incomes associated with a scheme are discounted to their present value for comparison, the analysis is termed a present value or discounted cash flow analysis. On the other hand if annual net incomes of different schemes are compared, this is termed the rate of return method. The latter is most frequently used by private organisations where tax returns and profits feature priminently. In such cases it is suggested that the assistance of qualified accountants is obtained. Present value comparisons are most common for public utilities.

A form of economic analysis popular in the United States is benefit/cost analysis. An economic benefit is attached to all products of a scheme, for instance a certain economic value is attached to water supplies, although this is difficult to evaluate in the case of domestic supplies. Those schemes with the highest benefit/cost values are attached highest priority. Where schemes are mutually exclusive such as is usually the case with public untilities the scheme with the largest present value of net benefit is adopted. It the total water requirements of a town for instance were fixed, the least-cost supply scheme would be selected for construction.

Uncertainty in Forecasts

Forecasts of demands, whether they be for water, oil or gas, are invariably clouded with uncertainty and risk. Strictly a probability analysis is required for each possible scheme, i.e. the net benefit of any particular scheme will be the sum of the net benefits multiplied by their probability for a number of possible demands. Berthouex (1971) recommends under-designing by 5 to 10% for pipelines to allow for uncertain forecasts, but his analysis does not account for cost inflation.

An alternative method of allowing for uncertainty is to adjust the discount or interest rate: increasing the rate will favour a low capital cost scheme, which would be preferable if the future demand were uncertain.

Example

A consumer requires 300 ℓ/s of water for 5 years then plans to increase his consumption to 600 ℓ/s for a further 25 years (the economic life of his factory).

He draws for 75% of the time every day. Determine the most economic diameter and the number of pipelines required. The water is supplied by a public body paying no tax. Power costs a flat 0.5 p/kWhr, which includes an allowance for operating and maintenance. The interest rate on loans (taken over 20 years) and on a sinking fund is 10% per annum, and the rate of inflation in cost of pipelines, pumps and power is 6% p.a. Pump and pumpstations costs amount to £300 per incremental kW, (including an allowance for standby plant) and pump efficiency is 70%.

The effective discount rate may be taken as the interest rate less the rate of inflation, i.e. 4% p.a., since £1 this year is worth £1 × 1.10/1.06 ≑ £1 × 1.04 next year.

The supply could be made through one large pipeline capable of handling 600 ℓ/s, or two smaller pipelines each delivering 300 ℓ/s, one installed five years after the other. A comparison of alternative diameters is made in the following table for a single pipeline. Similarly an analysis was made for two pipelines each delivering 300 ℓ/s. This indicated an optimum diameter of 600 mm for each pipeline and a total present value for both pipelines of £7 500 per 100 m. Thus one pipeline, 800 mm diameter, will be the most economic solution.

Note that the analysis is independent of the length of the pipeline, although it was assumed that pressure was such that a continuous low-pressure pipe was all that was required. Water hammer protection costs are assumed incorporated in the pipe cost here. The analysis is also independent of the capital loan period, although the results would be sensitive to change in the interest or inflation rates. Discount factors were obtained from present value tables for 4% over 5, 25 and 30 years periods. Uncertainty was not allowed for but would favour the two smaller pipelines.

Another interesting point emerged from the analysis: If a 600 mm pipeline was installed initially, due to a high uncertainty of the demand increasing from 300 to 600 ℓ/s, then if the demand did increase, it would be more economic to boost the pumping head and

Solution:

		600		700		800		900	
1.	Inside Dia.mm	600		700		800		900	
2.	Flow ℓ/s	300	600	300	600	300	600	300	600
3.	Head loss m/100 m	0.14	0.55	0.06	0.24	0.03	0.12	0.02	0.07
4.	Power loss kW/100 m = (3.)×Q/70	0.60	4.72	0.26	2.06	0.13	1.03	0.09	0.60
5.	Energy requirements kW hr/yr/ 100 m = (4.)×8760 × 0.75	3900	31000	1700	13500	850	6600	590	3900
6.	Annual pumping cost £/100 m	20	155	8	67	4	33	3	20
7.	Equiv.Capital cost of pumping over 5 years = (6.)×4.452	90	–	40	–	20	–	10	–
8.	Equiv.capital cost of pumping over 25 years. = (6.)×15.622	–	2430	–	1060	–	520	–	310
9.	Present value of pumping cost = (8.)/1.170		2070		900		440		260
10.	Cost of pumps etc. £/100 m:=(4.)×300	180	1410	80	620	40	310	30	180
11.	Present value of pump cost = (10.)/1.170 for second stage	180	1200	80	530	40	260	30	150
12.	Pipeline cost £/100 m	3600		4200		4800		5400	
13.	TOTAL COST £/100 m for 300 & 600 ℓ/s 7.+9.+11.+12	7140		5750		5560* (least cost)		5850	

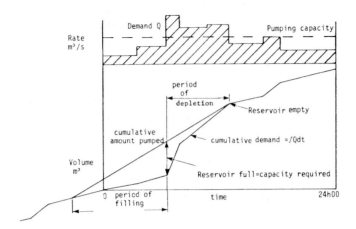

Fig. 1.4 Graphical calculation of reservoir capacity

pump the total flow through the one existing 600 mm diameter pipeline rather than provide a second 600 mm pipeline. This is indicated by a comparison of the present value of pumping through one 600 mm line (£7 140/100 m) with the present value of pumping through two 600 mm lines (£7 500/100 m).

BALANCING STORAGE

An aspect which deserves close attention in planning the pipeline system is reservoir storage. Demands such as those for domestic and industrial water fluctuate with the season, the day of the week and time of day. Peak-day demands are sometimes in excess of twice the mean annual demand whereas peak draw-off from reticulation systems may be six times the mean for a day. It would be uneconomic to provide pipeline capacity to meet the peak draw-off rates, and balancing reservoirs are normally constructed at the consumer end (at the head of the reticulation system) to meet these peaks. The storage capacity required varies inversely with the pipeline capacity.

The balancing storage requirement for any known draw-off pattern and pipeline capacity may be determined with a mass flow diagram:

Plot cumulative draw-off over a period versus time, and above this curve plot a line with slope equal to the discharge capacity of the pipeline. Move this line down till it just touches the mass draw-off period. Then the maximum ordinate between the two lines represents the balancing storage required (see Fig. 1.4).

An economic comparison is necessary to determine the optimum storage capacity for any particular system (Abramov, 1969) by adding the cost of reservoirs and pipelines and capitalizing running costs for different combinations and comparing them, the system with least total cost is selected. It is found that the most economic storage capacity varies from one day's supply based on the mean annual rate for short pipelines to two day's supply for long pipelines (over 60 km). Slightly more storage may be economic for small-bore pipelines (less than 450 mm diameter). In addition a certain amount of emergency reserve storage should be provided; up to 12 hours depending upon the availability of maintenance facilities.

REFERENCES

Abramov, N., 1969. Methods of reducing power consumption in pumping water. Int. Water Supply Assn. Congress, Vienna.
Berthouex, P.M., 1971. Accommodating uncertain forecasts. J. Am. Water Works Assn., 66 (1) 14.
Instn. of Civil Engs., 1962. An Introduction to Engineering Economics, London.
Osborne, J.M. and James, L.D., 1973. Marginal economics applied to pipeline design. Proc. Am. Soc. Civil Engs., 99 (TE3) 637.
White, J.E., 1969. Economics of large diameter liquid pipelines. Pipeline News, N.J.

LIST OF SYMBOLS

C – cost per unit of throughput

D – diameter

n – number of years

Q – throughput

r – interest rate on sinking fund

R – interest rate on loan

CHAPTER 2

HYDRAULICS

THE FUNDAMENTAL EQUATIONS OF FLUID FLOW

The three most important equations in fluid mechanics are the continuity equation, the momentum equation and the energy equation. For steady, incompressible, one-dimensional flow the continuity equation is simply obtained by equating the flow rate at any section to the flow rate at another section along the stream tube. By 'steady flow' is meant that there is no variation in velocity at any point with time. 'One-dimensional' flow implies that the flow is along a stream tube and there is no lateral flow across the boundaries of stream tubes. It also implies that the flow is irrotational.

The momentum equation stems from Newton's basic law of motion and states that the change in momentum flux between two sections equals the sum of the forces on the fluid causing the change. For steady, one-dimensional flow this is

$$\Delta F_x = \rho Q \Delta V_x \qquad (2.1)$$

where F is the force, ρ is the fluid mass density, Q is the volumetric flow rate, V is velocity and subscript x refers to the 'x' direction.

The basic energy equation is derived by equating the work done on an element of fluid by gravitational and pressure forces to the change in energy. Mechanical and heat energy transfer are excluded from the equation. In most systems there is energy loss due to friction and turbulence and a term is included in the equation to account for this. The resulting equation for steady flow of incompressible fluids is termed the Bernoulli equation and is conveniently written as:

$$\frac{V_1^2}{2g} + \frac{P_1}{\gamma} + Z_1 = \frac{V_2^2}{2g} + \frac{P_2}{\gamma} + Z_2 + h_\ell \qquad (2.2)$$

where V = mean velocity at a section

$V^2/2g$ = velocity head (units of length)

g = gravitational acceleration

p = pressure

p/γ = pressure head (units of length)

γ = unit weight of fluid

Z = elevation above an arbitrary datum

h_ℓ = head loss due to friction or turbulence between sections 1 & 2

The sum of the velocity head plus pressure head plus elevation is termed the total head.

Strictly the velocity head should be multiplied by a coefficient to account for the variation in velocity across the section of the conduit. The average value of the coefficient for turbulent flow is 1.06 and for laminar flow it is 2.0. Flow through a conduit is termed either uniform or non-uniform depending on whether or not there is a variation in the cross-sectional velocity distribution along the conduit.

For the Bernoulli equation to apply the flow should be steady, i.e. there should be no change in velocity at any point with time. The flow is assumed to be one-dimensional and irrotational. The fluid should be incompressible, although the equation may be applied to gases with reservations (Albertson et al., 1960).

The respective heads are illustrated in Fig. 2.1. For most practical cases the velocity head is small compared with the other heads, and it may be neglected.

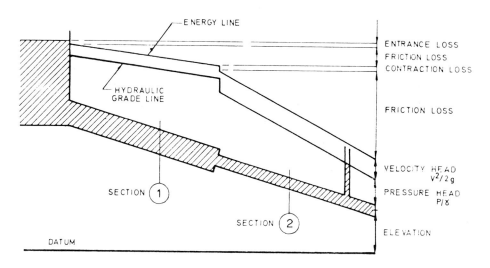

Fig. 2.1 Energy heads along a pipeline

FLOW HEAD LOSS RELATIONSHIPS

Empirical Flow Formulae

   The throughput or capacity of a pipe of fixed dimensions depends on the total head difference between the ends. This head is consumed by friction and other (minor) losses.

   The first friction head loss/flow relationships were derived from field observations. These empirical relationships are still popular in waterworks practice although more rational formulae have been developed. The head loss/flow formulae established thus are termed conventional formulae and are usually in an exponential form of the type

$$V = K \, R^x \, S^y \text{ or } S = K' Q^n / D^m$$

where V is the mean velocity of flow, K and K' are coefficients, R is the hydraulic radius (cross-sectional area of flow divided by the wetted perimeter, and for a circular pipe flowing full, equals one quarter of the diameter) and S is the head gradient (in m head loss per m length of pipe). Some of the equations more frequently applied are listed below:

	Basic Equation	SI units	f.p.s. units	
Hazen-Williams	$S = K_1 (V/C_w)^{1.85}/D^{1.167}$	$K_1 = 6.84$	$K_1 = 3.03$	(2.3)
Manning	$S = K_2 (nV)^2/D^{1.33}$	$K_2 = 6.32$	$K_2 = 2.86$	(2.4)
Chezy	$S = K_3 (V/C_z)^2/D$	$K_3 = 13.13$	$K_3 = 4.00$	(2.5)
Darcy	$S = \lambda V^2 / 2gD$	Dimension-less		(2.6)

   Except for the Darcy formula the above equations are not universal and the form of the equation depends on the units. It should be borne in mind that the formulae were derived for normal waterworks practice and take no account of variations in gravity, temperature or type of liquid. They are for turbulent flow in pipes over 50 mm diameter. The friction coefficients vary with pipe diameter, type of finish and age of pipe.

The conventional formulae are comparatively simple to use as they do not involve fluid viscosity. They may be solved directly as they do not require an initial estimate of Reynolds number to determine the friction factor (see next section). The rational equations cannot be solved directly for flow. Solution of the formulae for velocity, diameter or friction head gradient is simple with the aid of a slide rule, calculator, computer, nomograph or graphs plotted on log-log paper. The equations are of particular use for analysing flows in pipe networks where the flow/head loss equations have to be iteratively solved many times.

The most popular flow formula in waterworks practice is the Hazen-Williams formula. Friction coefficients for use in this equation are tabulated in Table 2.1. If the formula is to be used frequently, solution with the aid of a chart is the most efficient way. Many waterworks organizations use graphs of head loss gradient plotted against flow for various pipe diameters, and various C values. As the value of C decreases with age, type of pipe and properties of water, field tests are desirable for an accurate assessment of C.

TABLE 2.1   Hazen-Williams friction coefficients C

Type of Pipe	Condition			
	New	25 years old	50 years old	Badly corroded
PVC:	150	140	140	130
Smooth concrete, AC:	150	130	120	100
Steel, bitumen lined, galvanized:	150	130	100	60
Cast iron:	130	110	90	50
Riveted steel, vitrified woodstave:	120	100	80	45

For diameters less than 1 000 mm, subtract $0.1 \left(1 - \dfrac{Dmm}{1\ 000}\right) C$

Rational Flow Formulae

Although the conventional flow formulae are likely to remain in use for many years, more rational formulae are gradually gaining acceptance amongst engineers. The new formulae have a sound scientific basis backed by numerous measurements and they are universally applicable. Any consistent units of measurements may be used and liquids of various viscosities and temperatures conform to the proposed formulae.

The rational flow formulae for flow in pipes are similar to those for flow past bodies or over flat plates (Schlichting 1960). The original research was on small-bore pipes with artificial roughness. Lack of data on roughness for large pipes has been one deterrent to use of the relationships in waterworks practice.

The velocity in a full pipe varies from zero on the boundary to a maximum in the centre. Shear forces on the walls oppose the flow and a boundary layer is established with each annulus of fluid imparting a shear force onto an inner neighbouring concentric annulus. The resistance to relative motion of the fluid is termed kinematic viscosity, and in turbulent flow it is imparted by turbulent mixing with transfer of particles of different momentum between one layer and the next.

A boundary layer is established at the entrance to a conduit and this layer gradually expands until it reaches the centre. Beyond this point the flow becomes uniform. The length of pipe required for fully established flow is given by Schlichting, (1960).

$$\frac{x}{D} = 0.7\,Re^{1/4}\ \text{for turbulent flow.} \tag{2.7}$$

The Reynolds number $Re = VD/v$ is a dimensionless number incorporating the fluid viscosity v which is absent in the conventional flow formulae. Flow in a pipe is laminar for low Re (less than 2 000) and becomes turbulent for higher Re (normally the case in practice). The basic head loss equation is derived by setting the boundary shear force over a length of pipe equal to the loss in pressure multiplied by the area:

$$\tau \pi DL = \gamma h_f\ \ \pi D^2/4$$

$$\therefore h_f = \frac{4\,\tau/\gamma}{V^2/2g}\ \frac{L}{D}\ \frac{V^2}{2g}$$

$$= \lambda\frac{L}{D}\ \frac{V^2}{2g} \tag{2.8}$$

21

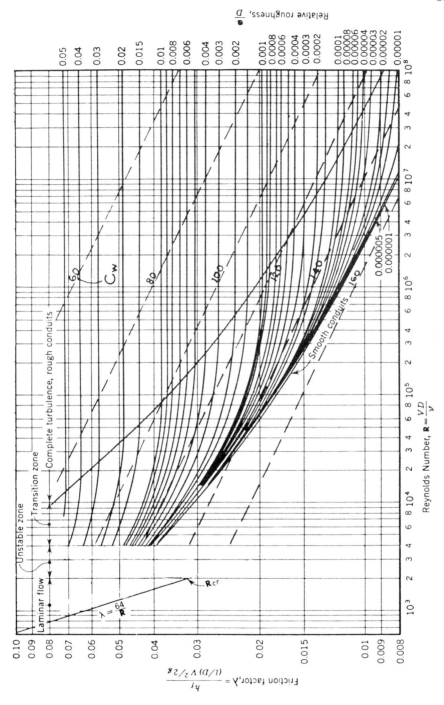

Fig. 2.2 Moody resistance diagram for uniform flow in conduits

where $\lambda = (4\tau/\gamma)(V^2/2g)$ (referred to as the Darcy friction factor), τ is the shear stress, D is the pipe diameter and h_f is the friction head loss over a length L. λ is a function of Re and the relative roughness e/D. For laminar flow, Poiseuille found that $\lambda = 64/Re$ i.e. λ is independent of the relative roughness. Laminar flow will not occur in normal engineering practice. The transition zone between laminar and turbulent flow is complex and undefined but is also of little interest in practice.

Turbulent flow conditions may occur with either a smooth or a rough boundary. The equations for the friction factor for both conditions are derived from the general equation for the velocity distribution in a turbulent boundary layer, which is derived from mixing length theory:

$$\tau = \rho k^2 \ell^2 (\frac{dv}{dy})^2$$

Integrating, with k = 0.4 and converting logs to base 10,

$$v/\sqrt{(\tau/\rho)} = 5.75 \log y/y' \tag{2.9}$$

where v is the velocity at a distance y from the boundary. For a hydrodynamically smooth boundary there is a laminar sub-layer, and Nikuradse found that $y' \propto v/\sqrt{\tau/\rho}$ so that

$$\frac{v}{\sqrt{\tau/\rho}} = 5.75 \log y \frac{\sqrt{\tau/\rho}}{v} + 5.5 \tag{2.10}$$

The constant 5.5 was found experimentally.

Where the boundary is rough the laminar sub-layer is affected and Nikuradse found that y' = e/30 where e is the boundary roughness.

Thus $\qquad \dfrac{v}{\sqrt{\tau/\rho}} = 5.75 \log \dfrac{y}{e} + 8.5 \tag{2.11}$

Re-arranging equations 2.10 and 2.11 and expressing v in terms of the average velocity V by means of the equation $Q = \int v dA$ we get

$$\frac{1}{\sqrt{\lambda}} = 2\log Re \sqrt{\lambda} - 0.8 \tag{2.12}$$

(turbulent boundary layer, smooth boundary) and

$$\frac{1}{\sqrt{\lambda}} = 2 \log \frac{D}{e} + 1.14 \tag{2.13}$$

(turbulent boundary layer, rough boundary)

Notice that for a smooth boundary, λ is independent of the relative roughness e/D and for a very rough boundary it is independent of the Reynolds number Re for all practical purposes.

Colebrook and White combined Equations 2.12 and 2.13 to produce an equation covering both smooth and rough boundaries as well as the transition zone:

$$\frac{1}{\sqrt{\lambda}} = 1.14 - 2 \log \left(\frac{e}{D} + \frac{9.35}{Re\sqrt{\lambda}}\right) \qquad (2.14)$$

Their equation reduces to Equation 2.12 for smooth pipes, and to Equation 2.13 for rough pipes. This semi-empirical equation yields satisfactory results for various commercially available pipes. Nikuradse's original experiments used sand as artificial boundary roughness. Natural roughness is evaluated according to the equivalent sand roughness. Table 2.2 gives values of e for various surfaces.

TABLE 2.2   Roughness of pipe materials   (Hydraulics Research station, 1969).

Value of e in mm for new, clean surface unless otherwise stated

Finish:	Smooth	Average	Rough
Glass, drawn metals	0	0.003	0.006
Steel, PVC or AC	0.015	0.03	0.06
Coated steel	0.03	0.06	0.15
Galvanized, vitrified clay	0.06	0.15	0.3
Cast iron or cement lined	0.15	0.3	0.6
Spun concrete or wood stave	0.3	0.6	1.5
Riveted steel	1.5	3	6
Foul sewers, tuberculated water mains	6	15	30
Unlined rock, earth	60	150	300

Fortunately λ is not very sensitive to the value of e assumed. e increases linearly with age for water pipes, the proportionality constant depending on local conditions. There may also be reduction in cross section with age.

The various rational fomulae for λ were plotted on a single graph by Moody and this graph is presented as Fig. 2.2. The kinematic viscosities of water at various temperatures are listed in the Appendix.

The Moody diagram is useful for calculating manually head loss if pipe velocity or flow rate is known. Unfortunately it is not very amenable to direct solution for the convrse, i.e. for velocity, given head loss, and a trial and error approach is necessary i.e. guess velocity to calculate Reynolds number, then read off λ , then recalculate velocity etc. Convergence is fairly rapid however.

The Colebrook White equation is easier to use if head loss is given and velocity is to be calculated. It is not so easy to solve for head loss or λ given velocity or flow however. The Hydraulics Research Station at Wallingford re-arranged the variables in the Colebrook-White equation to produce simple explicit flow/head loss graphs (Hydraulic Research Station, 1969):

Equation 2.14 may be arranged in the form

$$V = -2 \sqrt{2gDS} \, \log \left(\frac{e}{3.7D} + \frac{2.51\nu}{D \sqrt{2gDS}} \right) \tag{2.15}$$

Thus for any fluid at a certain temperature and defined roughness e, a graph may be plotted in terms of V, D and S. Fig. 2.3 is such a graph for water at 15°C and e = 0.06. The Hydraulic Research Station has plotted similar graphs for various conditions. The graphs are also available for non-circular sections, by replacing D by 4R. Going a step further, the Hydraulics Research Station re-wrote the Colebrook-White equation in terms of dimensionless parameters proportional to V, R and S, but including factors for viscosity, roughness and gravity. Using this form of the equation they produced a universal resistance diagram in dimensionless parameters. This graph was also published with their charts. Fig. 2.3 as an example is derived on a similar basis (Watson, 1979).

Comparison of Friction Formulae

Diskin (1960) presented a useful comparison of the friction factors from the Hazen-Williams and Darcy equations:

The Darcy equation may be written as

$$V = \sqrt{2g/\lambda} \, \sqrt{SD} \tag{2.16}$$

$$\text{or } V = C_z \sqrt{SR} \tag{2.17}$$

which is termed the Chezy equation and the Chezy coefficient is

$$C_z = \sqrt{8g/\lambda} \tag{2.18}$$

The Hazen-Williams equation may be rewritten for all practical purposes in the following dimensionless form:

$$S = 515(V/C_w)^2 \, (C_w/R_e)^{0.15}/gD \tag{2.19}$$

By comparing this with the Darcy-Weisbach equation (2.16) it may be deduced that

$$C_w = 42.4/(\lambda^{0.54} R_e^{0.08}) \tag{2.20}$$

25

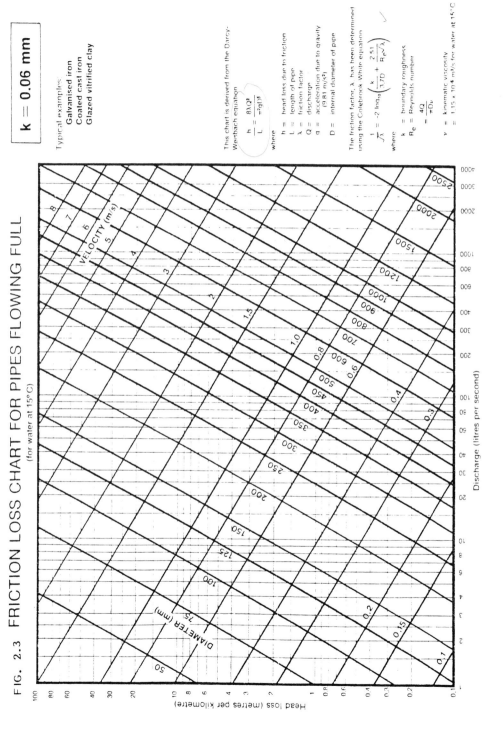

FIG. 2.3 FRICTION LOSS CHART FOR PIPES FLOWING FULL
(for water at 15°C)

The Hazen–Williams coefficient C_w is therefore a function of λ and R_e and values may be plotted on a Moody diagram (see Fig. 2.2). It will be observed from Fig. 2.2 that lines for constant Hazen–Williams coefficient coincide with the Colebrook–White lines only in the transition zone. In the completely turbulent zone for non–smooth pipes the coefficient will actually reduce the greater the Reynolds number i.e. one cannot associate a certain Hazen–Williams coefficient with a particular pipe as it varies depending on the flow rate. The Hazen–Williams equation should therefore be used with caution for high Reynolds numbers and rough pipes. It will also be noted that values of C_w above approximately 155 are impossible to attain in water–works practice (R_e around 10^6).

The Manning equation is widely used for open channel flow and part full pipes. The equation is

$$V = \frac{K}{n}R^{2/3}S^{\frac{1}{2}} \tag{2.21}$$

where K is 1.00 in SI units and 1.486 in ft lb units, and R is the hydraulic radius A/P where A is the cross–sectional area of flow and P the wetted perimeter. R is D/4 for a circular pipe, and in general for non–circular sections, 4R may be substituted for D.

TABLE 2.3   Values of Manning's 'n'

Smooth glass, plastic	0.010
Concrete, steel (bitumen lined)	
galvanized	0.011
Cast iron	0.012
Slimy or greasy sewers	0.013
Rivetted steel, vitrified wood–stave	0.015
Rough concrete	0.017

MINOR LOSSES

One method of expressing head loss through fittings and changes in section is the equivalent length method, often used when the conventional friction loss formulae are used. Modern practice is to express losses through fittings in terms of the velocity head i.e. $h_\ell = KV^2/2g$ where K is the loss coefficient. Table 2.4 gives typical loss

coefficients although valve manufacturers may also provide sup-
plementary data and loss coefficients K which will vary with gate
opening. The velocity V to use is normally the mean through the full
bore of the pipe or fitting.

TABLE 2.4   Loss coefficients for pipe fittings

Bends $h_B K_B V^2/2g$

Bend angle	Sharp	r/D=1	2	6
30°	0.16	0.07	0.07	0.06
45°	0.32	0.13	0.10	0.08
60°	0.68	0.18	0.12	0.08
90°	1.27	0.22	0.13	0.08
180°	2.2			
90° with guide vanes	0.2			

r   =   radius of bend to centre of pipe

A significant reduction in bend loss is possible if the radius is
flattened in the plane of the bend.

Valves $h_V = K_V V^2/2g$

Type:	Opening:	1/4	1/2	3/4	Full
Sluice		24	5.6	1.0	0.2
Butterfly		120	7.5	1.2	0.3
Globe					10 ₍
Needle		4	1	0.6	0.5
Reflux					1-2.5

Contractions and expansions in cross-section

Contractions:

$$h_c = K_c V_2^2/2g$$

Expansions

$$h_c = K_c V_1^2/2g$$

Wall-Wall		A_2/A_1						A_1/A_2				
Angle	0	0.2	0.4	0.6	0.8	1.0	0	0.2	0.4	0.6	0.8	1.0
7.5°							.13	.08	.05	.02	0	0
15°							.32	.24	.15	.08	.02	0
30°							.78	.45	.27	.13	.03	0
180°	.5	.37	.25	.15	.07	0	1.0	.64	.36	.17	.04	0

Entrance and exit losses: $h_e = K_e V^2/2g$

	Entrance	Exit
Protruding	0.8	1.0
Sharp	0.5	1.0
Bevelled	0.25	0.5
Rounded	0.05	0.2

PRESSURE AND FLOW CONTROL IN PIPES

Introduction

Valves and other fittings in pipes which reduce the head may be the cause of vapour bubble formation downstream. When the pressure is reduced water vaporizes. The bubbles may subsequently collapse giving rise to cavitation damage to the pipe or valve (Ball; 1970, Knapp et al, 1961). The geometry of the valve as well as the upstream and downstream pressure, degree of valve closure and flow rate affect the cavitation potential as well as the vapour pressure of the fluid. The conditions under which cavitation commences are referred to as critical. The cavitation damage increases as flow velocity increases and as the ratio of upstream to downstream pressure increases.

A number of empirical relationships for identifying the safe zone of operation of control valves have been proposed by valve suppliers. The popular measure of potential cavitation problem is the pressure reduction ratio, i.e. ratio of upstream to downstream pressures. Many commercially available valves are said to operate efficiently for pressure reducing ratios less than 3 or 4. It is however more logical to work with a cavitation index as indicated later. Since the geometry of the valve influences the critical conditions, the application of various types of valves for pressure reducing or flow control is described.

TYPES OF VALVES

Valves are used for two distinct purposes in pipelines, namely isolating or flow control.

Isolating Valves

Isolating valves are used to close off the flow through a pipe. They should be operated in the fully open or fully closed position only. They have poor flow characteristics in the partly open position. i.e. the obstruction to the flow is largely the exposed area of a gate and there is often little pressure reduction or flow control until the gate is practically shut. At that stage cavitation is possible and damage to the seat and valve body may result. This type of valve is not designed to reduce the pressure appreciably. i.e. it is designed to pass the full flow when fully open with minimal head loss.

Such valves include gate valves, butterfly valves and rotary or spherical valves. These valves are generally operated by hand and designed to close as rapidly as possible. Although gate valves are operated by spindles with screw threads, butterfly valves and rotary valves can be operated by turning a handle through 90° to fully close the valve from the fully open position. In order to overcome high seating pressures however and sometimes to control the rate of closure these valves may have reduction gear boxes in order to drive the spindle indirectly from the handwheel. Alternatively electrical and pneumatic control systems can be used to operate these valves.

Typical sluice valves, butterfly valves and spherical valves are illustrated in Figs. 10.1, 10.2 and 10.4. The relationship between open bore area and degree of turning of the handwheel is generally non linear. It can be observed in the case of the gate valve that the area reduces rapidly in the final stages of closure. This is unfavourable with respect to water hammer pressures as the velocity is reduced rapidly during the final stages of closure and this gives rise to higher water hammer pressures than the uniform rate of reduction in velocity in the pipe. The problems of cavitation are generally avoided with such valves as they are not operated in partly closed conditions except in emergencies.

Control Valves

Various types of valves have been designed to control the flow of water in pipelines by linearly reducing the open area and to contain the cavitation effect due to reduction of pressure through the valve. A typical valve in common use is illustrated in Fig. 10.3 (a needle valve). Diaphram type valves, slotted radial flow sleeve valves (Fig. 2.4) and plunger type control valves are also employed and there is a variation of the radial flow valve which has a piston covering the slots. Multiple expansion type valves and basket type high duty control valves (Fig. 2.5) are even more appropriate for pressure reducing. The former types rely on a single increase in velocity and any head dissipation due to expansion whereas the latter types which have variable resistance trim rings or layered baskets reduce the pressure in a number of stages and are said to suffer less cavitation damage as a result. This is largely because the pressure reducing ratio is limited at each stage but overall it may be relatively high. It is also recognised that these valves are less noisy when reducing high pressures.

The needle valve was once used widely in power station practice as it was recognised as a streamlined valve which could control flow at all stages of opening. It often had a pilot needle for final seating or the initial opening. The valve is however expensive owing to its construction and is now replaced to a large extent by simpler valves using sophisticated materials such as elastometers (the sleeve type valve) or by pistons or diaphrams in the case of low pressure valves (e.g. Dvir, 1981). The latter type of valve is not sensitive to poor quality water as the flexible diaphram prevents ingress of dirt into the workings. On the other hand the diaphram cannot accommodate high pressure differentials and suffers wear after prolonged operation. It may also be subject to vibration and instability as a venturi action is caused during the last stages of closing.

Piston type valves can be accurately controlled provided the flow is towards the exposed face of the piston. In some types the downstream face of the piston is exposed to the downstream pressure or the atmosphere and is supported by a spring. In others the downstream face of the piston can be subject to fluid pressure depending on the operation of a pilot control system. Such pilot systems can be operated hydraulically, that is off the water pressure. In this case strainers are often required to prevent the pilot tubes blocking. The system can also be operated off electrical solenoids or from a

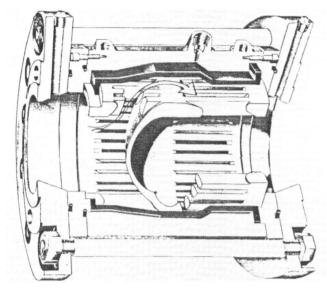

Fig. 2.4   Flexflo Sleeve valve for flow control

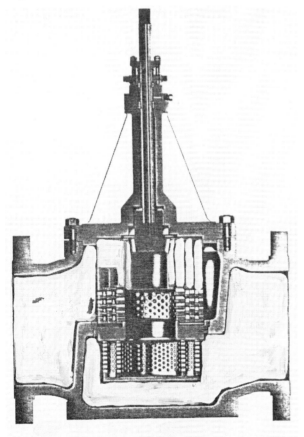

Fig. 2.5   High Duty Control valve with triple baskets for head
reduction

pneumatic support system.

The control system can be designed to limit the downstream pressure to a certain maximum figure. As the upstream pressure builds up and the sensor downstream senses an increasing pressure it will open a pilot valve held by means of a spring and this pilot valve will in turn permit a bigger pressure from upstream to come behind the piston and partly close the ports. Alternatively the valve may be used to control flow rate in which case the pressure on either side of an orifice or venturi is measured. The pressure difference in turn controls the pilot valve which will either permit a greater pressure to get behind the piston to close the valve or exhaust the pressure behind the piston to open the valve depending on which is required.

CAVITATION IN CONTROL VALVES

The mechanism whereby flow is reduced or pressure is reduced in a control valve is by converting the pressure energy upstream into a high velocity jet and then the velocity energy is dissipated in turbulence as it emerges into the open pipe again. The conversion from statical pressure energy to velocity energy is obtained from the Bernoulli equation;

$$\frac{P}{W} + Z + \frac{V^2}{2g} = \text{constant} \tag{2.22}$$

Here P is the upstream pressure, W is the unit weight of water so that P/W is the pressure head, Z is the elevation above a fixed datum and V is the water velocity. Thus if Z is a constant then the velocity possible assuming the pressure head is reduced to zero would be $V = \sqrt{2gh}$ where H = P/W. This however neglects upstream velocity and downstream pressure although the latter can be accounted for by replacing h by Δh. To account for upstream velocity head a coefficient is often introduced in the equation and in fact the pipe velocity is often given in the following form

$$v = C_d \sqrt{2g\Delta h} \tag{2.23}$$

where C_d is the discharge coefficient which may vary depending on the valve opening. Apart from the fact that the degree of opening of the valve affects the discharge coefficient, even if the discharge coefficient was separated from the degree of opening it would be found to vary owing to the different contraction ratio as a different area is open.

The actual stage of throttling at which cavitation commences can only be found from experiment. Cavitation has been classified as incipient i.e. the point at which vapour bubbles commenced to form, and choking i.e. the stage at which the discharge coefficient of the valve is appreciably affected by vaporization of the water downstream. The vaporization occurs because the pressure head is reduced and the energy is transformed into kinetic energy which impinges into a low pressure downstream. A number of parameters for assessing cavitation have been proposed (Winn and Johnson, 1970). The significant factor in the indication of cavitation is the cavitation index,

$$\sigma = \frac{P_d - P_v}{P_u - P_d} \qquad (2.24)$$

where P is the pressure and subscript d refers to downstream, u to upstream and v to vapour pressure.

For streamlined valves cavitation may not occur until the cavitation index drops as low as 0.1 whereas for poorly designed valves cavitation often occurs at a higher index i.e. nearer 1.0. In any case it will be found on inspection that most control valves experience a degree of cavitation at normal operating pressure ratios. Neglecting P_v it is seen from the cavitation index that if the critical index is 0.1 then the pressure can be reduced by a factor of nearly 10 whereas the pressure reduction ratio possible with unstreamlined valves is much lower. For an average critical cavitation index of 0.33 the pressure reducing ratio is from 2.24 neglecting vapour pressure, 4.0.

Various forms of the cavitation index have been proposed (Tullis, 1970). Very little is however known about the scaling down of the cavitation index and much work has to be done in this field yet.

The method of quantitatively assessing cavitation is also difficult. Various methods have been attempted i.e. estimation of the increase of velocity owing to the increase in bulk volume caused by bubble formation, photographic methods, pressure measurements due to collapse of the bubbles and sonic methods. Fig. 2.6 shows the variation in noise measured on a radial flow sleeve valve at different cavitation indices for a fixed degree of opening.

Interaction between Cavitation and Water Hammer Pressures

It is recognised that the collapse of vapour bubbles when the pressure subsequently increases may cause erosion of steel. This and the accompanying noise are often the symptoms of a poorly selected valve.

There is however an interesting relationship between the water hammer wave celerity and the cavitation potential. As a valve is closed so the bubble formation increases and this reduces the water hammer wave celerity downstream (Ch. 4 and 5).

If the free gas fraction can be related to the cavitation index which in turn is a function of the upstream or downstream pressure then it is possible to solve simultaneously the water hammer equation, the valve discharge equation and the cavitation index equation together with the equation relating gas release to cavitation index, for free gas volume downstream, downstream pressure and water hammer wave celerity. It is made difficult however by the fact that part of the cavity is not gas but vapour which rapidly turns into liquid when the pressure again increases. Small quantities of gas may be contained in the bubbles as they are released due to the drop in pressure. The bubbles may therefore not collapse completely. If the equations are solved (iteratively) then it will be found that the pressure downstream would initially drop rapidly as the valve is closed due to the water hammer deceleration effect. Then as bubbles begin to form the wave celerity reduces rapidly and the consequent pressure lowering is reduced so that the flow reduces more steadily. This study neglects wave reflections and so on but it does hold promise for the potential of cavitation to effect better control of flow as a valve closes especially during the initial stages of closure where this theory indicates that flow reduction is more rapid than was previously considered the case.

REFERENCES

Albertson, M.L., Barton, J.R. and Simons, D.B., 1960. Fluid Mechanics for Engineers. Prentice Hall, N.J.

Ball, J.W., 1970. Cavitation design criteria. In Tullis (1970). Proc. Inst. Colorado State Univ.

Diskin, M.H., Nov. 1960. The limits of applicability of the Hazen-Williams formulae. La Houille Blanche, 6.

Dvir, Y., 1981. Pressure regulators in water supply systems. Water and Irrigation Review, Water Works Association of Israel.

Hydraulics Research Station, 1969. Charts for the Hydraulic Designs of Channels and Pipes, 3rd Edn., H.M.S.O., London.

Knapp, R., Daily, J.W. and Hammitt, F.G.,1961. Cavitation. McGraw-Hill.

Schlichting, H., 1960. Boundary Layer Theory. 4th Edn., McGraw-Hill N.Y.

Tullis, J.P., 1970. Control of flow in closed conduits. Proc. Inst. Colorado State University.

Watson, M.D., July 1979. A simplified approach to the solution of pipe flow problems using the Colebrook-White method. Civil Eng. in S.A., 21(7), pp 169-171.

Winn, W.P. and Johnson, D.E. December 1970. Cavitation parameters for outlet valves. Proc. ASCE, HY12.

LIST OF SYMBOLS

A — cross-sectional area of flow

C — Hazen-Williams friction factor

C' — friction factor

C_z — Chezy friction factor

d — depth of water

D — diameter

e — Nikuradse roughness

f — Darcy friction factor (equivalent to λ)

F_x — force

g — gravitational acceleration

h_ℓ — head loss

h_f — friction head loss

K — loss coefficient

L — length of conduit

n — Manning friction factor

P — wetted perimeter

p — pressure

36

Q – flow rate

R – hydraulic radius

R_e – Reynolds number

S – hydraulic gradient

V – mean velocity across a section

v – velocity at a point

x – distance along conduit

y – distance from boundary

Z – elevation

γ – specific weight

ρ – mass density

τ – shear stress

ν – kinematic viscosity

λ – Darcy friction factor – (f in USA)

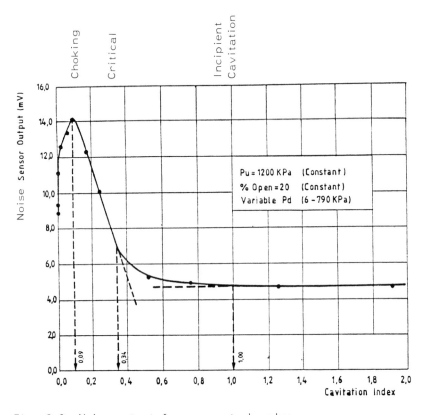

Fig. 2.6   Noise output from a control valve

CHAPTER 3

PIPELINE SYSTEM ANALYSIS AND DESIGN

NETWORK ANALYSIS

The flows through a system of interlinked pipes or networks are controlled by the difference between the pressure heads at the input points and the residual pressure heads at the drawoff points. A steady-state flow pattern will be established in a network such that the following two criteria are satisfied:-

(1) The net flow towards any junction or node is zero, i.e., inflow must equal outflow, and

(2) The net head loss around any closed loop is zero, i.e., only one head can exist at any point at any time.

The line head losses are usually the only significant head losses and most methods of analysis are based on this assumption. Head loss relationships for pipes are usually assumed to be of the form $h = K\ell Q^n/D^m$ where h is the head loss, ℓ is the pipe length, Q the flow and D the internal diameter of the pipe.

The calculations are simplified if the friction factor K can be assumed the same for all pipes in the network.

Equivalent Pipes for Pipes in Series or Parallel

It is often useful to know the equivalent pipe which would give the same head loss and flow as a number of interconnected pipes in series or parallel. The equivalent pipe may be used in place of the compound pipes to perform further flow calculations.

The equivalent diameter of a compound pipe composed of sections of different diameters and lengths in series may be calculated by equating the total head loss for any flow to the head loss through the equivalent pipe of length equal to the length of compound pipe:-

$$K(\Sigma\ell)Q^n/D_e^m = \Sigma K\ell Q^n/D^m$$

$$\therefore \qquad D_e = (\frac{\Sigma\ell}{\Sigma\ell/D^m})^{1/m} \qquad\qquad (3.1)$$

(m is 5 in the Darcy formula and 4.85 in the Hazen-Williams formula).

Similarly, the equivalent diameter of a system of pipes in <u>parallel</u> is derived by equating the total flow through the equivalent pipe 'e' to the sum of the flows through the individual pipes 'i' in parallel:

Now $h = h_i$

i.e. $K\ell_e Q^n / D_e^m = K\ell_i Q_i^n / D_i^m$

So $Q_i = (\ell_e/\ell_i)^{1/n} (D_i/D_e)^{m/n} Q$

and $Q = \Sigma Q_i = \Sigma [(\ell_e/\ell_i)^{1/n} (D_i/D_e)^{m/n} Q]$.

So cancelling out Q, and bringing D_e and ℓ_e to the left hand side,

$(D_e^m/\ell_e)^{1/n} = \Sigma (D_i^m/\ell_i)^{1/n}$

and if each ℓ is the same,

$$D_e = [\Sigma (D_i^{m/n})]^{n/m} \qquad (3.2)$$

The equivalent diameter could also be derived using a flow/head loss chart. For pipes in parallel, assume a reasonable head loss and read off the flow through each pipe from the chart. Read off the equivalent diameter which would give the total flow at the same head loss. For pipes in series, assume a reasonable flow and calculate the total head loss with assistance of the chart. Read off the equivalent pipe diameter which would discharge the assumed flow with the total head loss across its length.

It often speeds network analyses to simplify pipe networks as much as possible using equivalent diameters for minor pipes in series or parallel. Of course the methods of network analysis described below could always be used to analyse flows through compound pipes and this is in fact the preferred method for more complex systems than those discussed above.

Loop Flow Correction Method

The loop method and the node method of analysing pipe networks both involve successive approximations speeded by a mathematical technique developed by Hardy Cross (1936).

The steps in balancing the flows in a network by the loop method are:

(1) Draw the pipe network schematically to a clear scale. Indicate all inputs, drawoffs, fixed heads and booster pumps (if present).

(2) If there is more than one constant head node, connect pairs of constant head nodes or reservoirs by dummy pipes represented by dashed lines. Assume a diameter and length and calculate the flow corresponding to fixed head loss. In subsequent flow corrections, omit this pipe but include it in calculating head losses.

(3) Imagine the network as a pattern of closed loops in any order. To speed convergence of the solution some of the major pipes may be assumed to form large superimposed loops instead of assuming a series of loops side by side. Use only as many loops as are needed to ensure that each pipe is in at least one loop.

(4) Starting with any pipe assume a flow. Proceed around a loop containing the pipe, calculating the flow in each pipe by subtracting drawoffs and flows to other loops at nodes. Assume flows to other loops if unknown. Proceed to neighbouring loops one at a time, on a similar basis. It will be necessary to make as many assumptions as there are loops. The more accurate each assumption the speedier will be the solution.

(5) Calculate the head loss in each pipe in any loop using a formula such as $h = K\ell Q^n/D^m$ or use a flow/head loss chart (preferable if the analysis is to be done by hand).

(6) Calculate the net head loss around the loop, i.e., proceeding around the loop, add head losses and subtract head gains until arriving at the starting point. If the net head loss around the loop is not zero, correct the flows around the loop by adding the following increment in flow in the same direction that head losses were calculated:

$$\Delta Q = \frac{-\Sigma h}{\Sigma (hn/Q)} \qquad (3.3)$$

This equation is the first order approximation to the differential of the head loss equation and is derived as follows:-

Since $h = K\ell Q^n / D^m$

$dh = K\ell n Q^{n-1} dQ / D^m = (hn/Q)dQ$

Now the total head loss around each loop should be zero, i.e.,

$$\Sigma (h + dh) = 0$$
$$\Sigma h + \Sigma (hn/Q)dQ = 0$$
$$\Delta Q = \frac{- \Sigma h}{n \Sigma (h/Q)}$$

The value of h/Q is always positive (or zero if h and Q are zero).

(7)   If there is a booster pump in any loop, subtract the generated head from Σh before making the flow correction using the above equations.

(8)   The flow around each loop in turn is corrected thus (steps 5-7).

(9)   Steps 5-8 are repeated until the head around each loop balances to a satisfactory amount.

The Node Head Correction Method

With the node method, instead of assuming initial flows around loops, initial heads are assumed at each node. Heads at nodes are corrected by successive approximation in a similar manner to the way flows were corrected for the loop method. The steps in an analysis are as follows:-

(1)   Draw the pipe network schematically to a clear scale. Indicate all inputs, drawoffs, fixed heads and booster pumps.

(2)   Assume initial arbitrary heads at each node (except if the head at that node is fixed). The more accurate the initial assignments, the speedier will be the convergence of the solution.

(3)   Calculate the flow in each pipe to any node with a variable head using the formula $Q = (hD^m/K\ell)^{1/n}$ or using a flow/head loss chart.

(4)   Calculate the net inflow to the specific node and if this is not zero, correct the head by adding the amount

$$\Delta H = \frac{\Sigma Q}{\Sigma (Q/nh)} \qquad (3.4)$$

This equation is derived as follows:-

Since
$$Q = (hD^m/K\ell)^{1/n}$$

$$dQ = Qdh/nh$$

We require
$$\Sigma(Q + dQ) = 0$$

$$\Sigma Q + \frac{\Sigma Qdh}{nh} = 0$$

But
$$dH = -dh$$

So
$$\Delta H = \frac{\Sigma Q}{\Sigma(Q/nh)}$$

Flow Q and head loss h are considered positive if towards the node. H is the head at the node. Inputs (positive) and drawoffs (negative) at the node should be included in ΣQ.

(5) Correct the head at each variable-head node in similar manner, i.e. repeat steps 3 and 4 for each node.

(6) Repeat the procedure (steps 3 to 5) until all flows balance to a sufficient degree of accuracy. If the head difference between the ends of a pipe is zero at any stage, omit the pipe from the particular balancing operation.

Alternative Methods of Analysis

Both the loop method and the node method of balancing flows in networks can be done manually but a computer is preferred for large networks. If done manually, calculations should be set out well in tables or even on the pipework layout drawing if there is sufficient space. Fig. 3.1 is an example analysed manually by the node method. There are standard computer programs available for network analysis, most of which use the loop method.

The main advantage of the node method is that more iterations are required than for the loop method to achieve the same convergence, especially if the system is very unbalanced to start with. It is normally necessary for all pipes to have the same order of head loss. There are a number of methods for speeding the convergence. These include overcorrection in some cases, or using a second order approximation to the differentials for calculating corrections.

The node head correction method is slow to converge on account of

42

the fact that corrections dissipate through the network one pipe at a time. Also the head correction equation has an amplification factor (n) applied to the correction which causes overshoot.

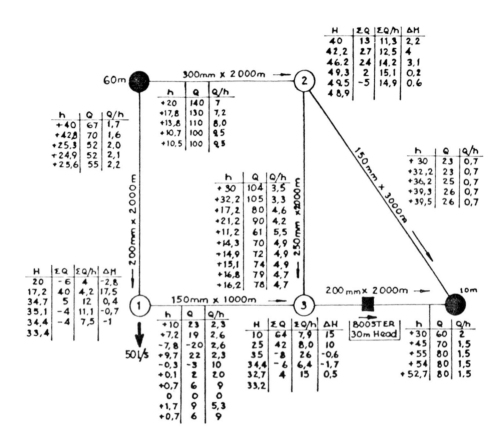

NOTES

Heads in metres, flows in litres per second, diameters in millimetres, lengths in metres. Arrows indicate positive direction of h & Q (arbitrary assumption). Blackened circles indicate nodes with fixed heads, numbers in circles indicate order in which nodes were corrected. Head losses evaluated from Fig. 2.3.

$$H = \frac{1.85 \quad Q \text{ in}}{Q/h}$$

Fig. 3.1   Example of node method of network flow analysis

The loop flow correction method has the disadvantage in data preparation. Flows must be assumed around loops, and drawoffs are defined indirectly in the assumed pipe flows. The added effort in data preparation and interpretation often offsets the quicker convergence. This is so because of relatively low computing costs compared with data assembly. Trial and error design is also cumbersome if loop flows have to be changed each time a new pipe is added.

Network Analysis by Linear Theory

The Hardy Cross methods of network analysis are suited to manual methods of solution but suffer drawbacks in the effort required in comparison with computer orientated numerical methods. The latter involve the simultaneous solution of sets of equations describing flow and head balance. Simultaneous solution has the effect that very few iterations are required to balance a network when compared with the number of iterations for the loop flow correction and node head correction methods. On the other hand solution of a large number of simultaneous equations, even if rendered linear, requires a large computer memory and many iterations.

Newton–Raphson techniques for successive approximation of non-linear equations are mathematically sophisticated but the engineering problem becomes subordinate to the mathematics. Thus Wood and Charles (1972) linearized the head loss equation, improving the linear approximation at each step and establishing equations for head balance around loops.

Isaac and Mills (1980) similarly linearized the head loss equation as follows for flow between nodes i and j:

$$Q_{ij} = C_{ij}^{\frac{1}{2}}(H_i - H_j) / \sqrt{|H_i - H_j|} \qquad (3.5)$$

where the term in the square root sign is assumed a constant for each iteration. If the Darcy friction equation is employed,

$$C_{ij} = (\pi/4)^2 \, 2g \, D^5 / \lambda L$$

Substitute equation 3.5 into the equation for flow balance at each node:

$$\sum_i Q_{ij} = Q_j$$

where Q_j is the drawoff at node j and Q_{ij} is the flow from node i to node j, negative if from j to i. There is one such equation for each node.

If each Q_{ij} is replaced by the linearized expression in 3.5, one has a set of simultaneous equations (one for each node) which can be solved for H at each node. The procedure is to estimate H at each node initially, then solve for new H's. The procedure is repeated until satisfactory convergence is obtained.

OPTIMIZATION OF PIPELINE SYSTEMS

The previous section described methods for calculating the flows in pipe networks with or without closed loops. For any particular pipe network layout and diameters, the flow pattern corresponding to fixed drawoffs or inputs at various nodes could be calculated. To design a new network to meet certain drawoffs, it would be necessary to compare a number of possibilities. A proposed layout would be analysed and if corresponding flows were just sufficient to meet demands and pressures were satisfactory, the layout would be acceptable. If not, it would be necessary to try alternative diameters for pipe sizes and analysis of flows is repeated until a satisfactory solution is at hand. This trial and error process would then be repeated for another possible layout. Each of the final networks so derived would then have to be costed and that network with least cost selected.

A technique of determining the least-cost network directly, without recourse to trial and error, would be desirable. No direct and positive technique is possible for general optimization of networks with closed loops. The problem is that the relationship between pipe diameters, flows, head losses and costs is not linear and most routine mathematical optimization techniques require linear relationships. There are a number of situations where mathematical optimization techniques can be used to optimize layouts and these cases are discussed and described below. The cases are normally confined to single mains or tree-like networks for which the flow in each branch is known. To optimize a network with closed loops, random search techniques or successive approximation techniques are needed.

Mathematical optimization techniques are also known as systems analysis techniques (which is an incorrect nomenclature as they are design techniques not analysis techniques), or operations research techniques (again a name not really descriptive). The name mathematical optimization techniques will be retained here. Such techniques include simulation (or mathematical modelling) coupled with a selection technique such as steepest path ascent or random searching.

The direct optimization methods include dynamic programming, which is useful for optimizing a series of events or things, transportation programming, which is useful for allocating sources to demands and linear programming, for inequalities (van der Veen, 1967 and Dantzig, 1963). Linear programming usually requires the use of a computer, but there are standard optimization programs available.

Dynamic Programming for Optimizing Compound Pipes

One of the simplest optimization techniques, and indeed one which can normally be used without recourse to computers, is dynamic programming. the technique is in fact only a systematic way of selecting an optimum program from a series of events and does not involve any mathematics. The technique may be used to select the most economic diameters of a compound pipe which may vary in diameter along its length depending on pressures and flows. For instance, consider a trunk main supplying a number of consumers from a reservoir. The diameters of the trunk main may be reduced as drawoff takes place along the line. The problem is to select the most economic diameter for each section of pipe.

A simple example demonstrates the use of the technique. Consider the pipeline in Fig. 3.2. Two consumers draw water from the pipeline, and the head at each drawoff point is not to drop below 5 m, neither should the hydraulic grade line drop below the pipe profile at any point. The elevations of each point and the lengths of each section of pipe are indicated. The cost of pipe is £0.1 per mm diameter per m of pipe. (In this case the cost is assumed to be independent of the pressure head, although it is simple to take account of such a variation). The analysis will be started at the downstream end of the pipe (point A). The most economic arrangement will be with minimum residual head i.e. 5 m, at point A. The head, H, at point B may be anything between 13 m and 31 m above the datum, but to simplify the analysis, we will only consider three possible heads with 5 m increments between them at points B and C.

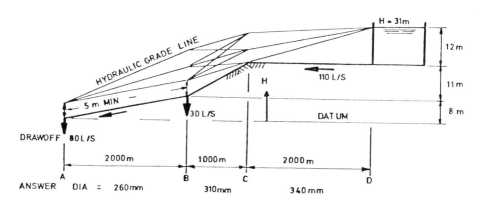

Fig. 3.2   Profile of pipeline optimized by dynamic programming

The diameter D of the pipe between A and B, corresponding to each of the three allowed heads may be determined from a head loss chart such as Fig. 2.3 and is indicated in Table 3.1 (I) along with the corresponding cost.

We will also consider only three possible heads at point C. The number of possible hydraulic grade lines between B and C is 3 × 3 = 9, but one of these is at an adverse gradient so may be disregarded. In Table 3.1 (II) a set of figures is presented for each possible hydraulic grade line between B and C. Thus if H_B = 13 and H_C = 19 then the hydraulic gradient from C to B is 0.006 and the diameter required for a flow of 110 ℓ/s is 310 mm (from Fig. 2.3). The cost of this pipeline would be 0.1 × 310 × 1 000 = £31000. Now to this cost must be added the cost of the pipe between A and B, in this case £60 000 (from Table 3.1 (I)). For each possible head

TABLE 3.1 Dynamic programming optimization of a compound pipe

I

HEAD AT B H_B	HYDR. GRAD. h_{B-A}	DIA. mm D_{B-A}	COST D_{B-A} COST £
13	.004	300	60000
18	.0065	260	52000
23	.009	250	50000

II

$H_C=$	19			24			29		
H_B	h_{C-B}	D_{C-B}	COST £	h_{C-B}	D_{C-B}	COST £	h_{C-B}	D_{C-B}	COST £
13	.006	310	31000 60000 91000*	.011	270	27000 60000 87000	.016	250	25000 60000 85000
18	.001	430	43000 52000 95000	.006	310	31000 52000 83000*	.011	270	27000 52000 79000*
23	–	–	–	.001	430	43000 50000 93000	.006	310	31000 50000 81000

III

H_C	h_{D-C}	D_{D-C}	COST £
19	.006	310	62000 91000 153000
24	.0035	340	68000 83000 151000*
29	.001	430	86000 79000 165000

H_C there is one minimum total cost of pipe between A and C, marked with an asterisk. It is this cost and the corresponding diameters only which need be recalled when proceeding to the next section of pipe. In this example, the next section between C and D is the last and there is only one possible head at D, namely the reservoir level.

In Table 3.1 (III) the hydraulic gradients and corresponding diameters and costs for Section C – D are indicated. To the costs of pipe for this section are added the costs of the optimum pipe arrangement up to C. This is done for each possible head at C, and the least total cost selected from Table 3.1 (III). Thus the minimum possible total cost is £151 000 and the most economic diameters are 260, 310 and 340 mm for Sections A – B, B – C and C – D respectively. It may be desirable to keep pipes to standard diameters in which case the nearest standard diameter could be selected for each section as the calculations proceed or each length could be made up of two sections; one with the next larger standard diameter and one with the next smaller standard diameter, but with the same total head loss as the theoretical result.

Of course many more sections of pipe could be considered and the accuracy would be increased by considering more possible heads at each section. The cost of the pipes could be varied with presssures. A booster pump station could be considered at any point. in which case its cost and capitalized power cost should be added in the tables. A computer may prove useful if many possibilities are to be considered, and there are standard dynamic programming programs available.

It will be seen that the technique of dynamic programming reduces the number of possibilities to be considered by selecting the least–cost arrangement at each step. Kally (1969) and Buras and Schweig (1969) describe applications of the technique to similar and other problems.

Transportation programming for least–cost allocation of resources

Transportation programming is another technique which normally does not required the use of a computer. The technique is of use primarily for allocating the yield of a number of sources to a number of consumers such that a least–cost system is achieved. The cost of delivering the resource along each route should be linearly

49

proportional to the throughput along that route and for this reason the technique is probably of no use in selecting the optimum pipe sizes. It is of use, however, in selecting a least-cost pumping pattern through an existing pipe distribution system, provided the friction head is small in comparison with static head, or for obtaining a planning guide before demands are accurately known.

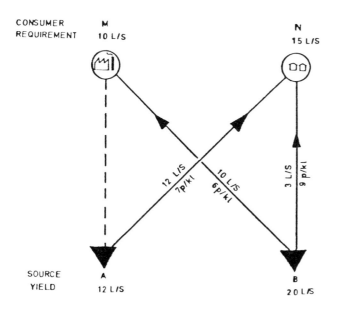

Fig. 3.3 Least-cost allocation pattern for transportation programming example

An example serves to illustrate the technique. In this example, there are two sources of water, A and B, and two consumers, M and N. A and B could deliver 12 and 20 ℓ/s respectively and M and N require 10 and 15 ℓ/s respectively. thus there is a surplus of water. The cost of pumping along routes A – M, A – N, B – M and B – N are 5, 7, 6 and 9 p/1 000 litres respectively.

The data are set out in tabular form for solution in Tabel 3.2 (1). Each row represents a source and each column a demand. The unit cost of delivery along each route is indicated in the top right corner of the corresponding block in the table. The first step is to make an arbitrary initial assignment of resources in such a manner that each yield and demand is satisfied. Starting with the top left block of the table, the maximum possible allocation is 10. This satisfies the demand of column M and the amount is written in the bottom left corner of block AM. Proceeding to the next column, since the first column is completed, the maximum possible allocation in the first row is 2, which satisfies the yield of row A. So the next block to be considered is in row B, namely column N. Proceed through the table making the maximum possible assignment at each stage until all resources are allocated (even if to the slack column). Thus the next allocation is the 13 in the second row, then the 7 in the third column.

TABLE 3.2   Transportation programming optimization of an allocation

(I)

SOURCE YIELD	CONSUMER:	M	N	SURPLUS	EVALUATION
	REQUIRE- MENTS:	10	15	7	NUMBER:
		5	7	0	0
A 12		10	2	-2	
		6	9	0	2
B 20		7 | 13	7		
EVALUATION NUMBER:		5	7	-2	

(II)

		10	15	7	
		5	7	0	0
A 12		4 | 12		-2	
		6	9	0	2
B 20		10	3	7	
		4	7	-2	

Once an initial allocation is made the figures are re-arranged methodically until a least-cost distribution emerges. To decide which would be the most profitable arrangement, assign a relative evalua-tion number to each row and column as follows:-

Assign the value 0 to row 1 and work out the other evaluation numbers such that the sum of the row evaluation number and column evaluation number is equal to the cost coefficient for any occupied block. The value for column M is 5, for column N is 7, for row B is 2, and so on. Now write the sum of the row and column evaluation numbers beneath the cost coefficient of each unoccupied block If this sum if bigger than the cost coefficient of the block, it would pay to introduce a resource allocation into the block. This is not easy to see immediately, but stems from the method of determining each evaluation sum from the cost coefficients of occupied blocks. The biggest possible rate of improvement is indicated by the biggest difference between the evaluation sum and the cost coefficient. The biggest and in fact in our case the only, improvement would be to introduce an amount into Block BM. The maximum amount which can be put in block BM is determined by drawing a closed loop using occupied blocks as corners (see the dotted circuit in Table 3.2 (I)). Now for each unit which is added to block BM, one unit would have to be subtracted from block BN, added to block AN and subtracted from block AM to keep the yields and requirements consistent. In this case the maximum allocation to BM is 10, since this would evacuate block AM. The maximum re-distribution i.e. 10 is made, and the amount in the block at each corner of the closed loop adjusted by 10 to satisfy yields and requirements. Only one re-distribution of resources should be done at a time.

After making the best new allocation, re-calculate the evaluation number and evaluation sums as in Table 3.2 (II). Allocate resource to the most profitable block and repeat the re-distribution procedure until there is no further possible cost improvement, indicated by the fact that there is no evaluation sum greater than the cost coefficient in any block. In our example we arrived at the optimum distribution in two steps, but more complicated patterns involving more sources and consumers may need many more attempts.

The example can only serve to introduce the subject of transportation programming. There are many other conditions which are dealt with in textbooks on the subject of mathematical optimization techniques such as Van der Veen (1967) and Dantzig (1963) and this example only serves as an introduction. For instance, if two blocks in the table happened to be evacuated simultaneously, one of the blocks could be allocated a very small quantity denoted by 'e' say. Computations then proceed as before and the quantity 'e' disregarded at the end.

Linear Programming for Design of Least-Cost Open Networks

Linear programming is one of the most powerful optimization techniques. The use of a computer is normally essential for complex systems, although the simple example given here is done by hand. The technique may only be used if the relationship between variables is linear, so it is restricted in application. Linear programming cannot be used for optimizing the design of pipe networks with closed loops without resort to successive approximations. It can be used to design trunk mains or tree-like networks where the flow in each branch is known. Since the relationships between flow, head loss, diameter and cost are non-linear, the following technique is used to render the system linear: For each branch or main pipe, a number of pre-selected diameters are allowed and the length of each pipe of different diameter is treated as the variable. The head losses and costs are linearly proportional to the respective pipe lengths. The program will indicate that some diameter pipes have zero lengths, thereby in effect eliminating them. Any other type of linear constraint can be treated in the analysis. It may be required to maintain the pressure at certain points in the network above a fixed minimum (a linear inequality of the greater-than-or-equal-to-type) or within a certain range. The total length of pipe of a certain diameter may be restricted because there is insufficient pipe available.

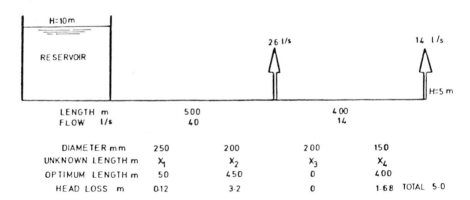

Fig. 3.4 Least-cost trunk main by linear programming

The example concerns a trunk main with two drawoff points (Fig.

3.4). The permissible diameters of the first leg are 250 and 200 mm, and of the second leg, 200 and 150 mm. There are thus four variables, X_1, X_2, X_3 and X_4 which are the lengths of pipe of different diameters. This simple example could be optimized by manual comparison of the costs of all alternatives giving the correct head loss, but linear programming is used here to demonstrate the technique.

The head losses per 100 m of pipe and costs per m for the various pipes are indicated below:-

Diameter mm	Head loss @ 40 ℓ/s m/100 m	@ 14 ℓ/s m/100 m	Cost £100/100 m
250	0.25		5
200	0.71	0.1	4
150		0.42	3

The linear constraints on the system are expressed in equation form below and the coefficients of the equations are tabulated in Table 3.3 (I). Lengths are expressed in hundred metres.

Lengths

$$X_1 + X_2 = 5$$
$$X_3 + X_4 = 4$$

Head Loss:
$$0.25X_1 + 0.71X_2 + 0.1X_3 + 0.42X_4 = 5$$

Objective Function:
$$5X_1 + 4X_2 + 4X_3 + 3X_4 = minimum$$

The computations proceed by setting all real variables to zero, so it is necessary to introduce artificial slack variables into each equation to satisfy the equality. The slack variables are designated a, b and c in Table 3.3 (I), and their cost coefficients are set at very high values designated m. To initiate the solution, the slack variables a, b and c are assigned the values 5, 4 and 5 respectively (see the third column of Table 3.3 (I)),

The numbers in any particular line of the main body of the table indicate the amount of the program variable which would be displaced by introducing one unit of the column variable. Thus one unit of X_1 would displace 1 unit of a and 0.25 units of c.

To determine whether it is worthwhile replacing any variable in the program by any other variable, a number known as the oppor-

TABLE 3.3   Linear Programming Solution of Pipe Problems

I

Prog. Variable	Cost Coeff.	Amount	X_1 5	X_2 4	X_3 4	X_4 3	a m	b m	c m	Repl. ratio
a	m	5	1	1			1			5/1*
b	m	4			1	1		1		∞
c	m	5	0.25	0.71	0.10	0.42			1	5/.71
OPPORTUNITY VALUE:			5-1.25m	4-1.71m	4-1.1m	3-1.42m	0	0	0	*key row
				KEY COLUMN						

II

			X_1 5	X_2 4	X_3 4	X_4 3	a m	b m	c m	
X_2	4	5	1	1			1			∞
b	m	4			1	1		1		4
c	m	1.45	-0.46		0.1	0.42	-0.71		1	3.45*
			1+0.46	0	4-1.1m	3-1.42m	1.71m-4	0	0	

III

			X_1 5	X_2 4	X_3 4	X_4 3	a m	b m	c m	
X_2	4	5	1	1			1			5
b	m	0.55	1.1		0.76		1.69		-2.38	0.5*
X_4	3	3.45	-1.1		0.24	1	-1.69		2.38	
			11-1.1m	0	3.28-0.76m	0	1.1-0.69m		3.38-8.2	

IV

			X_1 5	X_2 4	X_3 4	X_4 3	a m	b m	c m	
X_2	4	4.5		1	-0.69		1		2.16	
X_1	5	0.5	1		0.69		1.52		-2.16	
X_4	3	4			1	1				
			0	0	0.31	0	m-	m-	m-	

(No further improvement possible)

tunity number is calculated for each column. If one unit of X_1 was introduced, then the cost would increase by (5 - (1 × m) - (0 × m) - (0.25 × m)), which is designated the opportunity value, i.e. the opportunity value for each column is calculated by multiplying the entries in that column by the corresponding cost coefficients of the program variable in the second column and subtracting the total thus formed from the cost coefficient of the column variable. The most profitable variable to introduce would be X_2, since it shows the greatest cost reduction per unit (or negative opportunity value). The X_2 column is now designated the key column. The key column is that which shows the lowest opportunity value (in the cost mini-mization case). Only one variable may be introduced at a time.

To determine the maximum amount of the key column variable which may be introduced, calculate the replacement ratios for each row as follows:-

Divide the amount of the program variable for each row by the corresponding number in the key column. The lowest positive replace-ment ratio is selected as that is the maximum amount which could be introduced without violating any of the constraints. The row with the lowest positive replacement ratio is designated the key row and the number at the intersection of the key column and key row, the key number.

After introducing a new variable, the matrix is rearranged (Table 3.3 (II)) so that the replacement ratios remain correct. The program variable and its cost coefficient in the key row are replaced by the new variable and its cost coefficient. The amount column as well as the body of the table are revised as follows:-

Each number in the key row is divided by the key number.

From each number in a non-key row, subtract the corresponding number in the key row multiplied by the ratio of the old row number in the key column divided by the key number. The new tableau is given as Table 3.3 (II).

The procedure of studying opportunity values and replacement ratios and revising the table is repeated until there is no further negative opportunity value. In the example Table 3.3 (IV) shows all positive opportunity values so the least-cost solution is at hand (indicated by the current program variables and their corresponding values).

The reader should refer to a standard textbook on linear programming (e.g. Van der Veen, (1967) and Dantzig (1963)) for a full description of the technique. There are many other cases which can only be mentioned below:-

(1) If the constraints are of the $\leq$ (less-than-or-equal-to) type and not just equations, slack variables with zero cost coefficients are introduced into the l.h.s. of each constraint to make them equations. The artificial slack variables with high cost coefficients are then omitted.

(2) If the constraints are of the $\geq$ (greater-than-or-equal-to) type, introduce artificial slack variables with high cost coefficients into the l.h.s. of the constraint and subtract slack variables with zero cost coefficients from each inequality to make them equations.

(3) If the objective function is to be minimized, the opportunity value with the highest negative value is selected, but if the function is to be maximized, the opportunity value with the highest positive value is selected.

(4) The opportunity values represent shadow values of the corresponding variables i.e. they indicate the value of introducing one unit of that variable into the program.

(5) If two replacement ratios are equal, whichever row is selected, the amount of program variable in the other row will be zero when the matrix is rearranged. Merely assume it to have a very small value and proceed as before.

An example of two stage optimization, namely first the layout and then the pipe diameters is given in Stephenson (1984). Non-linear programming and search methods are also discussed. For instance Lam (1973) described a gradient optimization method.

REFERENCES

Buras, N. and Schweig, Z., 1969. Aqueduct route optimization by dynamic programming. Proc. Am. Soc. Civil Engrs. 95 (HY5).
Cross, H., 1936. Analysis of flow in networks of conduits or conductors. University of Illinois Bulletin 286.
Dantzig, G.B., 1963. Linear Programming and Extensions, Princeton University Press, Princeton.
Isaacs, L.T. and Mills, K.G., 1980. Linear theory methods for pipe networks analysis. Proc. Am. Soc. Civil Engrs. 106 (HY7).
Kally, E., 1969. Pipeline planning by dynamic computer programming. J. Am. Water Works Assn., (3).
Lam, C.F., 1973. Discrete gradient optimization of water systems. Proc. Am. Soc. Civil Engrs., 99 (HY6).
Stephenson, D., 1984. Pipeflow Analysis, Elsevier, 204 pp.
Van der Veen, B., 1967. Introduction to the Theory of Operational Research. Cleaver Hume, London.
Wood, D.J. and Charles, O.A., 1972. Network analysis using linear theory. Proc. ASCE, 98 (HY7) p 1157 – 1170.

LIST OF SYMBOLS

C – cost

D – diameter

h – head loss

H – head

K – constant

ℓ – length

Q – flow

58

CHAPTER 4

WATER HAMMER AND SURGE

RIGID WATER COLUMN SURGE THEORY

Transient pressures caused by a change of flow rate in conduits are often the cause of bursts. The pressure fluctuations associated with sudden flow stoppage can be several hundred metres head.

Transients in closed conduits are normally classed into two categories: Slow motion mass oscillation of the fluid is referred to a surge, whereas rapid change in flow accompanied by elastic strain of the fluid and conduit is referred to as water hammer. For slow or small changes in flow rate or pressure the two theories yield the same results.

It is normally easier to analyse a system by rigid column theory (whenever the theory is applicable) than by elastic theory. With rigid column theory the water in the conduit is treated as an incompressible mass. A pressure difference applied across the ends of the column produces an instantaneous acceleration. The basic equation relating the head difference between the ends of the water column in a uniform bore conduit to the rate of change in velocity is derived from Newton's basic law of motion, and is

$$h = \frac{-L}{g} \frac{dv}{dt} \tag{4.1}$$

where h is the difference in head between the two ends, L is the conduit length, v is the flow velocity, g is gravitational acceleration and t is time.

The equation is useful for calculating the head rise associated with slow deceleration of a water column. It may be used for calculating the water level variations in a surge shaft following power trip or starting up in a pumping line, or power load changes in a hydro-electric installation fed by a pressure pipeline. The equation may be solved in steps of Δt by computer, in tabular form or graphically. The following example demonstrates the numerical method of solution of the equation:-

Example

A 100 m long penstock with a cross sectional area, A_1, of 1 m^2 is protected against water hammer by a surge shaft at the turbine, with a cross sectional area, A_2, of 2 m^2 and an unrestricted orifice. The initial velocity in the conduit is 1 m/s and there is a sudden complete load rejection at the turbine. Calculate the maximum rise in water level in the surge shaft neglecting friction.

Take Δt = 1 sec. Then from Equ. 4.1, Δv = $-gh\Delta t/L$ = $-9.8h/100$ = $-0.098h$. By continuity, Δh = $A_1 v \Delta t/A_2$ = $1v/2$ = $0.5v$.

t	Δh = 0.5v	h	Δv = $-0.098h$	v
0.1	0.5	0.5	-0.049	0.951
1-2	0.476	0.976	-0.096	0.855
2-3	0.428	1.404	-0.138	0.717
3-4	0.359	1.763	-0.173	0.544
4-5	0.272	2.035	-0.199	0.345
5-6	0.172	2.207	-0.216	0.129
6-7	0.064	2.271*	-0.223	0.094

The maximum rise is 2.27 m, which may be compared with the analytical solution obtained from Equ. 4.16, of 2.26 m. The accuracy of the numerical method could be improved by taking smaller time intervals or taking the mean v and h over the time intervals to calculate Δh and Δv respectively. The method can readiiy be extended to include head losses, and is calculator-orientated.

Another useful application of the rigid water column equation is with water column separation. Following the stopping of a pump at the upstream end of a pumping line, the pressure frequently drops sufficiently to cause vaporization at peaks along the line. In such cases the water column beyond the vapour pocket will decelerate slowly and rigid column theory is sufficiently accurate for analysis. Equ. 4.1 may be integrated twice with respect to time t to determine the distance the water column will travel before stopping. If the pumps stop instantaneously the volume of the vapour pocket behind the water column of length ℓ will be Q = $A\ell v_o^2/2gh$ where v_o is the initial flow velocity.

MECHANICS OF WATER HAMMER

Water hammer occurs, as the name implies, when a column of water is rapidly decelerated (or accelerated). The rigid water column theory would indicate an infinitely large head rise if a valve in a pipeline carrying liquid were slammed shut instantaneously. There is, however, a certain amount of relief under these conditions, due to the elasticity of the fluid and of the pipe itself. Thus if a valve at the discharge end of a pipeline is shut the fluid upstream of the gate will pack against the gate, causing a pressure rise. The pressure will rise sufficiently to stop the liquid in accordance with the momentum law. The amount of water stopped per unit time depends on the amount of water required to replace the volume created by the compression of water and expansion of the pipe.

It can thus be shown that the relationship between the head rise Δh, the reduction in velocity Δv and the rate of progress of the wave front is $\Delta h = -c\Delta v/g$ (4.2a)

where c is the wave celerity and g is gravitational acceleration. This equation is often referred to as Joukowsky's law. It can further be proved from a mass balance that

$$c = 1/\sqrt{\frac{w}{g}\left(\frac{1}{K} + \frac{kd}{Ey}\right)} \qquad (4.3a)$$

If the pipe has a rigid lining (elastic modulus E_2 and thickness y_2) and rigid side fill (modulus E_s) the equation to use is

$$c = 1/\sqrt{\frac{w}{g}\left(\frac{1}{K} + \frac{d}{E_1y_1 + E_2y_2 + E_sd/5}\right)} \qquad (4.3b)$$

where w is the unit weight of liquid, K is its bulk modulus, d is the pipe diameter, E is its elastic modulus, y the pipe wall thickness and k a factor which depends on the end fixity of the pipe (normally about 0.9). c may be as high as 1370 m/s for a rigid walled tunnel or as low as 850 m/s for thin wall steel pipes. In the case of plastic pipe, or when free air is present (see Chapter 5), c may be as low as 200 m/s.

The pressure wave caused by the valve closure referred to earlier, thus travels upstream, superimposed on the static head, as illustrated in Fig. 4.1(1). When the wave front reaches the open reservoir end, the pressure in the pipe forces water backwards into the reservoir, so that the velocity now reverses and pressure drops back to static (reservoir) pressure again.

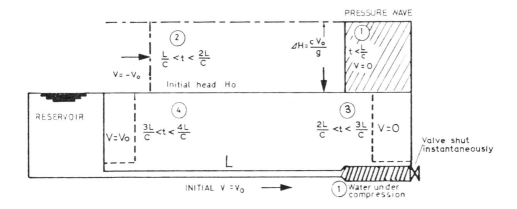

Fig. 4.1   Water hammer wave at different stages

A negative wave (2) thus travels downstream from the reservoir. That wave front will in turn reach the closed end. Now the velocity in the entire pipeline is $-v_o$ where v_o was the original velocity. The negative wave of head amplitude cv_o/g below static, travels back up the pipe to the reservoir. Upon reaching the reservoir it sucks water into the pipe, so that the velocity in the pipe reverts to $+v_o$ and the head reverts to static head. The sequence of waves will repeat itself indefinitely unless damped by friction.

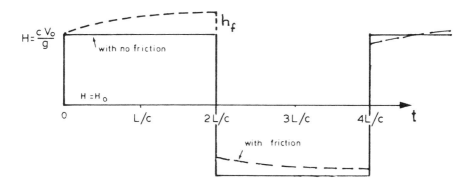

Fig. 4.2   Head fluctuations at valve end.

The variation in head at the valve will be as indicated in Fig. 4.2. In the case of pumping lines, the most violent change in flow conditions is normally associated with a pump trip. The water downstream of the pump is suddenly decelerated, resulting in a sudden reduction in pressure. The negative wave travels towards the discharge end (Fig. 4.3), where the velocity reverses and a positive wave returns towards the pumps. The pressure at the pump alternates from a pressure drop to a pressure rise.

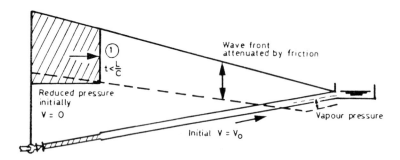

Fig. 4.3 Water hammer head drop after pump trip

The wave is complicated by line friction, changes in cross section, vaporization, or gradual flow variation. Thus if there is a negative pressure, the water will vaporize and air will be drawn in via air valves and from solution. The effect of friction can most readily be seen from Fig. 4.4 which illustrates a wave at different stages as it travels up the pipeline from a rapidly shut valve. The pressure heads behind the wave are not quite horizontal, due to the 'packing' effect causing some flow across the wave front.

The effect of changes in section or branch pipes can be included in one equation. The wave head change Δh^1 after reaching a junction is related to the original head change Δh^0 by the equation

$$\Delta h^1 = \frac{2 \Delta h^0 A_1/c_1}{A_1/c_1 + A_2/c_2 + A_3/c_3 + \dots} \qquad (4.4a)$$

$$\doteq 2 \Delta h^0 / (1 + A_2/A_1 + A_3/A_1 + \dots) \qquad (4.4b)$$

where A_i is the cross sectional area of pipe i and $c_1 \doteq c_2 \doteq c_3$.

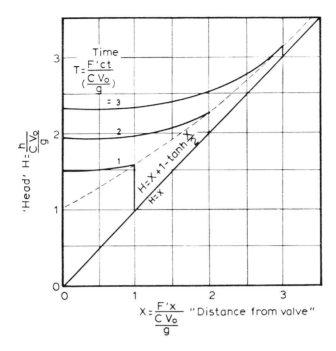

Fig. 4.4 Head at points along along pipeline with friction, for instantaneous stoppage

When a valve in a gravity main is closed gradually, the effect is analogous to a series of minute waves emanating from the valve. The system could be analysed numerically or graphically. The graphical form of analysis is useful for demonstrating the principles of simultaneous solution of the valve discharge equation and the water hammer equation. In fact the lines drawn on a graph are very similar to the so-called characteristic lines adopted in numerical solutions.

Fig. 4.5 illustrates a graphical analysis of a pipeline with a valve at the discharge end. The valve is closed over a period equal to 4 L/c and the valve discharge characteristics for four different degrees of closure are plotted on the graph. The line relating friction head loss to pipe velocity is also indicated.

To compute the head at the valve at time L/c after initiating closure one applies the water hammer equation 4.2 between points R and S. Thus

$$\Delta h = -(c/g) \, \Delta v - \Delta h_f \qquad\qquad (4.2b)$$

where Δh_f is the difference in frction head between R and S. Similarly applying the water hammer equation from S to R,

$$\Delta h = + (c/g) \, \Delta v + h_f \qquad\qquad (4.2c)$$

Computations thus proceed along lines of slope + or $-$ c/g. Ultimately the waves peak and then die out due to friction. The maximum head at point S occurs at t = 4 L/c and is 182 metres according to the computations on Fig. 4.5.

ELASTIC WATER HAMMER THEORY

The fundamental differential wave equations relating pressure to velocity in a conduit may be derived from consideration of Newton's law of motion and the conservation of mass respectively, and are:

$$\frac{\partial h}{\partial x} + \frac{1}{g} \frac{\partial v}{\partial t} + \frac{\lambda v |v|}{2gd} = 0 \qquad\qquad (4.5)$$

$$\frac{\partial h}{\partial t} + \frac{c^2}{g} \frac{\partial v}{\partial x} = 0 \qquad\qquad (4.6)$$

The last term in Equ. 4.5 accounts for friction which is assumed to obey Darcy's equation with a constant friction factor. The assumption of a steady-state friction factor for transient conditions is not strictly correct. Tests indicate that head losses during transient flow are higher than those predicted using the friction factor applicable to normal flow conditions. Energy is probably absorbed during flow reversals when the velocity is low and the friction factor consequently relatively high.

Methods of Analysis

A common method of analysis of pipe systems for water hammer pressures used to be graphically (Lupton, 1953). Fig. 4.6 is such a chart for the maximum and minimum heads at the downstream valve for no line friction. The valve area is assumed to reduce linearly to zero over time T and the valve discharge coefficient is assumed constant. To use the chart calculate the valve closure parameter cT/L

65

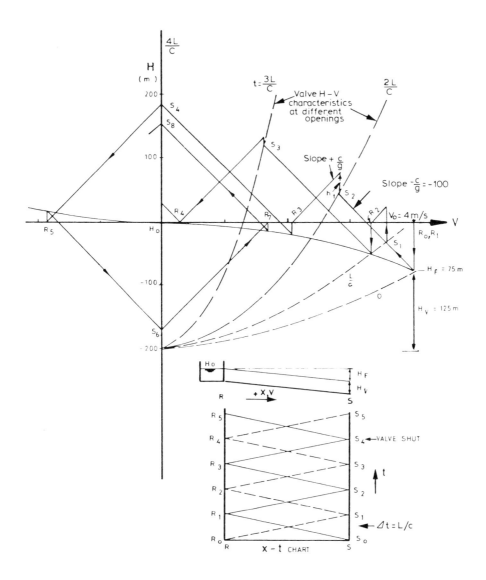

Fig. 4.5   Graphical water hammer analysis for slow valve
closure with friction

and valve head loss parameter $h_\ell/(cv_o/g)$. Read off on the vertical axis the maximum head parameter $h'/(cv_o/g)$ (full lines) and minimum head parameter $-h/(cv_o/g)$ (dashed lines). Multiply by cv_o/g and the answers are the maximum and minimum heads respectively above and below static head. Note that the chart is for linear reduction in area with time, a condition rarely encountered in practice. The characteristics of sluice and butterfly valves are such that most of the flow reduction occurs at the end of the valve stroke, with the result that the water hammer heads are higher than predicted by the chart. A more accurate analysis is therefore necessary for important lines.

The most economical method of solution of the water hammer equations for particular systems is by digital computer. Solution is usually by the method of characteristics (Streeter and Lai, 1963 and Streeter and Wylie, 1950) which differs little in principle from the old graphical method. The differential water hammer equations are expressed in finite difference form and solved for successive time intervals. The conduit is divided into a number of intervals and Δt is set equal to $\Delta x/c$. The $x - t$ grid on which solution takes place is depicted in Fig. 4.7. Starting from known conditions along the pipeline at time t, one proceeds to calculate the head and velocity at each point along the line at time $t + \Delta t$.

By expressing equations 4.5 and 4.6 as total differentials and adding, one gets two simultaneous equations involving dh and dv:

$$\text{For } \frac{dx}{dt} = + c \; : \; dh + \frac{c}{g}dv + \frac{c\lambda v|v|dt}{2gd} = 0 \tag{4.7a}$$

$$\text{For } \frac{dx}{dt} = - c \; : \; dh - \frac{c}{g}dv - \frac{c\lambda v|v|dt}{2gd} = 0 \tag{4.7b}$$

Equs. 4.7a and 4.7b may be solved for h_p' and v_p' at point p at time $t + \Delta t$ in terms of known h and v at two other points q and r at time t:

$$h_p' = \frac{h_q + h_r}{2} + \frac{c}{g}\frac{v_q - v_r}{2} + \frac{c\lambda dt}{2gd}\frac{v_r|v_r| - v_q|v_q|}{2} \tag{4.8a}$$

$$v_p' = \frac{v_q + v_r}{2} + \frac{g}{c}\frac{h_q - h_r}{2} - \frac{\lambda dt}{2d}\frac{v_r|v_r| + v_q|v_q|}{2} \tag{4.8b}$$

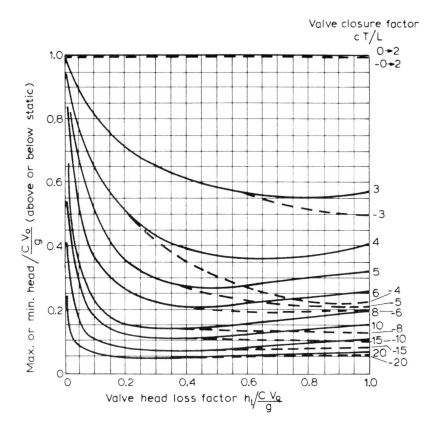

Fig. 4.6 Maximum and minimum head at downstream valve for linear closure

At the terminal points, an additional condition is usually imposed; either h is fixed, or v is a function of a gate opening or pump speed. The correct Equ. 4.7a or 4.7b is solved simultaneously with the known condition to evaluate the new h and v at time t + Δt. The computations commence at known conditions and are terminated when the pressure fluctuations are sufficiently damped by friction.

Where a branch pipe s occurs or there is a change in diameter, then Equs. 4.8a and 4.8b should be replaced by Equs. 4.9 and 4.10:

$$h'_p = [h_q A_q + h_r A_r + h_s A_s + (c/g)(q_q - q_r - q_s) - (c\lambda dt/2g)(q_q |q_q|/d_q A_q -$$

$$q_r |q_r|/d_r A_r - q_s |q_s|/A_s)] / (A_q + A_r + A_s) \qquad (4.9)$$

then

$$q'_{qp} = q_{qp} + (A_q g/c)(h_q - h'_p) - \lambda q_q |q_q| dt/2d_q A_q \qquad (4.10a)$$

$$q'_{pr} = q_{pr} + (A_r g/c)(h'_p - h_r) - \lambda q_r |q_r| dt/2d_r A_r \qquad (4.10b)$$

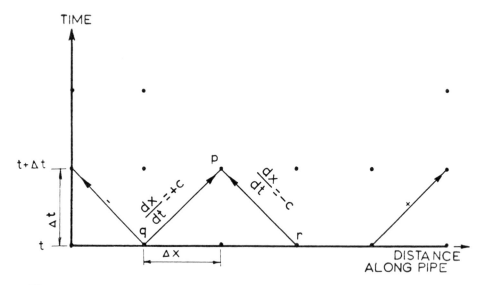

Fig. 4.7 x – t Grid for water hammer analysis by characteristic method.

$$q'_{ps} = q_{ps} + (A_s g/c)(h'_p - h_s) - \lambda q_s |q_s| dt / 2d_s A_s \qquad (4.10c)$$

Effect of Friction

Fluid friction damps the water hammer waves as they travel along the conduit. If there is no exciting influence the waves will gradually die away and the pressure along the conduit will tend to static pressure.

The characteristic method of solution by computer accurately predicts the effect of friction provided there is no discontinuity in the wave. At a sharp wave front, it is necessary to resort to some other method of analysis. Fortunately an analytical solution is feasible at the wave front. Ludwig (1950) demonstrated that the amplitude of a water hammer wave travelling back along a line with friction following instantaneous stoppage is indicated by the hyperbolic function (see Fig. 4.4):

$$h = cv_o/g \left[1 - \tanh \left(\frac{\lambda x v_o^2}{4gd} \middle/ \frac{cv_o}{g} \right) \right] \qquad (4.11)$$

In pumping lines following sudden pump stopping the maximum over-pressures at the pump will exceed the pumping head if the friction head is greater than approximately $0.7 cv_o/g$. Fig. 4.8 indicates minimum and maximum head envelopes along pipelines with various friction heads following instantaneous stoppage at the up-stream end. To use the chart, multiply the ordinates by cv_o/g and plot the maximum and minimum head envelopes for the correct friction factor $h_f/(cv_o/g)$, above or below static on a pipe profile drawing.

The chart is only valid provided there is no column separation when the negative wave travels up the pipeline i.e. the minimum head envelope should at no point fall below 10 m below the pipeline profile. It has, however, been found from experience that the maximum heads with column separation are often similar to those without separation.

Fig. 4.8 may also be used to determine the maximum and minimum heads along pipelines with friction following sudden flow stoppage at the downstream end by a closing valve. Turn the chart upside down and read off maximum envelope instead of minimum and vice- versa.

PROTECTION OF PUMPING LINES (Stephenson, 1972)

The pressure transients following power failure to electric motor driven pumps are usually the most extreme that a pumping system will experience. Nevertheless, the over-pressures caused by starting pumps should also be checked. Pumps with steep head/flow characteristics often induce high over-pressures when the power is switched on. This is because the flow is small (or zero) when the pump is switched on so a wave with a head equal to the closed valve head is generated. By partly closing the pump delivery valves during starting, the over-pressures can be reduced.

The over-pressures caused by closing line valves or scour valves should also be considered.

If the pumps supplying an unprotected pipeline are stopped sud-denly, the flow will also stop. If the pipeline profile is relatively close to the hydraulic grade line, the sudden deceleration of the water column may cause the pressure to drop to a value less than atmospheric pressure. The lowest value to which pressure could drop is vapour pressure. Vaporization or even water column separation may thus occur at peaks along the pipeline. When the pressure wave is returned as a positive wave the water columns will rejoin giving rise to water hammer over-pressures.

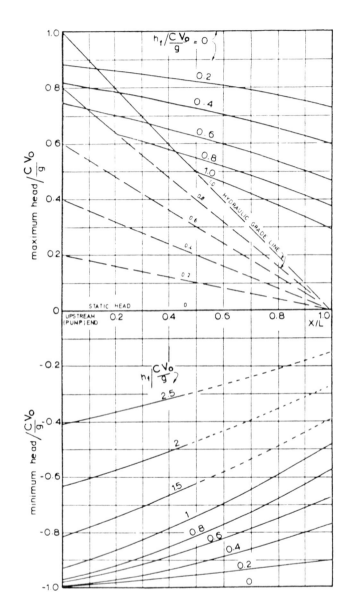

Fig. 4.8   Maximum and minimum head envelopes following
instantaneous pump stopping in pipelines with
friction

Unless some method of water hammer protection is installed, a pumping pipeline system will normally have to be designed for a water hammer overhead equal to cv_o/g. In fact this is often done with high-pressure lines where water hammer heads may be small in comparison with the pumping head. For short lines this may be the most economic solution, and even if water hammer protection is installed it may be prudent to check that the ultimate strength of the pipeline is sufficient should the protective device fail.

The philosophy behind the design of most methods of protection against water hammer is similar. The object in most cases is to reduce the downsurge in the pipeline caused by stopping the pumps. The upsurge will then be correspondingly reduced, or may even be entirely eliminated. The most common method of limiting the down-surge is to feed water into the pipe as soon as the pressure tends to drop.

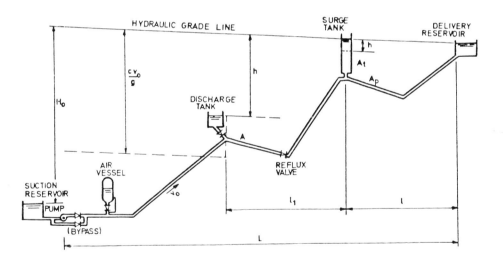

Fig. 4.9   Pipeline profile illustrating suitable locations for various devices for water hammer protection.

Suitable locations for various protective devices are illustrated in Fig. 4.9. Most of the systems involve feeding water into the pipe.

Observe that in all cases the sudden momentum change of the water column beyond the tank is prevented so the elastic water hammer phenomenon is converted to a slow motion surge phenomenon. Part of the original kinetic energy of the water column is converted into potential energy instead of elastic energy. The water column gradually decelerates under the effect of the difference in heads between the ends. If it was allowed to decelerate the water column would gather momentum in the reverse direction and impact against the pump to cause water hammer over-pressures. If, however, the water column is arrested at its point of maximum potential energy, which coincides with the point of minimum kinetic energy, there will be no sudden change in momentum and consequently no water hammer over-pressure. The reverse flow may be stopped by installing a reflux valve or throttling device at the entrance to the discharge tank or air vessel, or in the pipeline. A small orifice bypass to the reflux valve would then allow the pressures on either side to gradually equalize.

Fortunately charts are available for the design of air vessels and for investigation of the pump inertia effects, so that a water hammer analysis is not normally necessary. Rigid water column theory may be employed for 'the analysis of surge tank action, and in some cases, of discharge tanks.

If the pipeline system incorporates in-line reflux valves or a pump bypass valve, an elastic water hammer analysis is usually necessary. The analysis may be done graphically or, if a number of solutions of similar systems are envisaged, a computer program could be developed. Normally the location, size and discharge characteristics of a protective device such as a discharge tank have to be determined by trial and error. The location and size of in-line or bypass reflux valves may similarly have to be determined by trial. In these instances a computer program is usually the most economical method of solution, as a general program could be developed, and by varying the design parameters methodically, an optimum solution arrived at.

The following sections describe various methods of reducing water hammer in pumping lines, and offer design aids.

Pump Inertia

If the rotational inertia of a centrifugal pump and motor continue to rotate the pump for a while after power failure, water hammer pressure transients may be reduced. The rotating pump, motor and entrained water will continue to feed water into the potential vacuum on the delivery side, thereby alleviating the sudden deceleration of the water column. The effect is most noticeable on low-head, short pipelines.

After the power supply to the motor is cut off, the pump will gradually slow down until it can no longer deliver water against the delivery head existing at the time. If the delivery head is still higher than the suction head, it will then force water through the pump in the reverse direction, with the pump still spinning in the forward direction, provided there is no reflux or control valve on the delivery side of the pump. The pump will rapidly decelerate and gather momentum in the reverse direction, and will act as a turbine under these conditions. The reverse speed of the pump will increase until it reaches runaway speed. Under these conditions there is a rapid deceleration of the reverse flow and water hammer over-pressures will result.

If there is a reflux valve on the delivery side of the pump, the reverse flow will be arrested, but water hammer overpressures will still occur. The pressure changes at the pump following power failure may be calculated graphically (Parmakian, 1963) or by computer.

The pump speed N after a time increment Δt is obtained by equating the work done in decelerating the pump to the energy transferred to the water:

$$\tfrac{1}{2}M \left[\left(\frac{2\pi N_1}{60}\right)^2 - \left(\frac{2\pi N_2}{60}\right)^2 \right] F_N = wHQ\Delta t$$

Solving for $N_1 - N_2 = \Delta N$ we get

$$\Delta N = \frac{900wHQ\Delta t}{\pi^2 MNF_N} \qquad\qquad (4.12)$$

where M is the moment of inertia of the pump impeller and motor, N is the speed in rpm, F_N is the pump efficiency at the beginning of the time interval, w is the unit weight of fluid, H is the pumping head above suction head and q is the discharge. As an approximation, $F_N = (N/N_o)F_o$ where subscript o refers to initial conditions.

The head/discharge characteristics of the pump can be fairly accurately represented by the equation

$$H = aN^2 + bNQ - cQ^2 \tag{4.13}$$

a, b and c are constants which can be evaluated provided three points on the head/discharge/speed characteristics of the pump are known.

Substituting q = vA where A is the cross sectional area of the pipe, and using Equ. 4.8, two equations result involving two unknowns which may be solved for h and v at the pump.

Fortunately the results of a large number of analyses have been summarized in chart form. Charts presented by Parmakian indicate the maximum downsurge and upsurge at the pumps and midway along the pipeline for various values of pump inertia and the pipeline parameter P. The charts also indicate the maximum reverse speed and the times of flow reversal, zero pump speed and maximum reverse speed.

Kinno and Kennedy (1965) have prepared comprehensive charts which include the effect of line friction and pump efficiency; specific speed of the pump is also discussed. Of interest is a chart indicating maximum upsurge if reverse rotation of the pump is prevented. Kinno suggests that the upsurge could be reduced considerably if reverse flow through the pump was permitted, but with reverse rotation prevented.

Fig. 4.10 summarizes the maximum and minimum pressures which will occur in various pumping systems following power failure. Values for the minimum head are taken from Kinno's paper, and the maximum head values are adjusted to cover typical pump efficiencies around 85 per cent. It will be observed that the minimum head often occurs when the first wave returns to the pump (t = 2L/c). The values of the subsequent upsurge at the pump if flow reversal is permitted, are also presented in the chart. These values are lower than the magnitude of the upsurge which would occur if reverse flow was prevented with a reflux valve. If flow reversal was prevented the maximum headrise above operating head H_o would be approximately equal to the lowest head-drop below H_o.

The author has a simple rule of thumb for ascertaining whether the pump inertia will have an effect in reducing the water hammer pressures. If the inertia parameter $I = MN^2/wALH_o^2$ exceeds 0.01, the pump inertia may reduce the downsurge by at least 10 per cent. Here M is the moment of inertia of the pump, N is the speed in rpm and AL is the volume of water in the pipe. The expression was derived from Fig. 4.10, and it was assumed that c/g is approximately equal to 100.

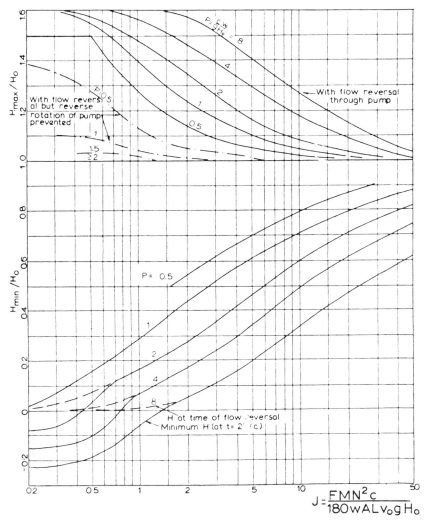

Fig. 4.10   Maximum and minimum heads at pump after power failure

Some installations have a fly-wheel fitted to the pump to increase the moment of inertia. In most cases the flywheel would have to be impractically heavy, also it should be borne in mind that starting currents may thereby be increased.

In subsequent sections the effect of pump inertia is neglected and the pumps are assumed to stop instantaneously.

Pump Bypass Reflux Valve

One of the simples arrangements of protecting a pumping line against water hammer is a reflux valve installed in parallel with the pump (Fig. 4.11). The reflux or non-return valve would discharge only in the same direction as the pumps. Under normal pumping conditions the pumping head would be higher than the suction head and the pressure difference would maintain the reflux valve in a closed position. On stopping the pumps, the head in the delivery pipe would tend to drop by the amount cv_o/g. The head may drop to below the suction head, in which case water would be drawn through the bypass valve. The pressure would therefore only drop to the suction pressure less any friction loss in the bypass. The return wave overpressure would be reduced accordingly.

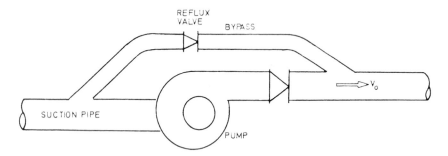

Fig. 4.11 Pump with bypass reflux valve.

This method of water hammer protection cannot be used in all cases, as the delivery pressure will often never drop below the suction pressure. In other cases there may still be an appreciable water hammer overpressure (equal in value to the initial drop in pressure). The method really has use only when the pumping head is considerably less than cv_o/g. In addition, the initial drop in pressure along the entire pipeline length should be tolerable. The suction reservoir level should also be relatively high or there may still be column separation in the delivery line.

Normally the intake pipes draw directly from a constant head reservoir. However, there may be cases where the intake pipe is fairly long and water hammer could be a problem in it too. In these cases a bypass reflux valve would, in a similar way to that described above, prevent the suction pressure exceeding the delivery pressure.

It should be noted that water may also be drawn through the pump during the period that the delivery head is below the suction head, especially if the machine was designed for high specific speeds, as is the case with through-flow pumps. In some cases the bypass reflux valve could even be omitted, although there is normally a fairly high head loss through a stationary pump.

Surge Tanks

The water surface in a surge tank is exposed to atmospheric pressure, while the bottom of the tank is open to the pipeline. The tank acts as a balancing tank for the flow variations which may occur, discharging in case of a head drop in the pipe, or filling in case of a head rise. Surge tanks are used principally at the head of turbine penstocks, although there are cases where they can be applied in pumping systems. It is seldom that the hydraulic grade line of a pumping line is low enough to enable an open tank to be used. It may be possible to construct a surge tank at a peak in the pipeline profile and protect the pipeline between the pumps and the tank against water hammer by some other means. If the surge tank is relatively large, it could be treated as the discharge end of the intermediate pipeline length, and this section could be treated as an independent pipeline shorter in length than the original pipeline.

The fluctuations of the water surface level in a surge tank following power failure may be studied analytically. The transients between the pumps and surge tank are high-frequency water hammer phenomena which will not affect the water level in the tank noticeably. The analysis concerns the slow motion surges of the water column between the surge tank and delivery end of the line.

Rigid water column theory may be used in the study, since elastic water hammer waves will not pass the open tank. The rate of deceleration of the water column is $dv/dt = -gh/\ell$ (4.14) where h is the height of the delivery head above the level of the surge tank and ℓ is the pipeline length between the tank and delivery end. The head h gradually increases as the level in the tank drops (see Fig. 4.9).

Another equation relating h and v may be derived by considering continuity of flow at the tank outlet:

$$A_p v = A_t dh/t \qquad\qquad (4.15)$$

where A_p and A_t are the cross sectional areas of the pipe and tank respectively. Note that the flow from the pump side is assumed to be zero.

Solving Equs. 4.14 and 4.15 and using the fact that at $t = 0$, $h = 0$ and $v = v_o$ to evaluate the constants of integration, one obtains an expression for h:

$$h = v_0 \sqrt{A_p \ell / A_t g} \ \sin \ (t\sqrt{A_p g / A_t \ell}) \qquad\qquad (4.16)$$

From this equation it is apparent that the amplitude of the down-surge in the tank is $v_o \sqrt{A_p \ell / A_t g}$ and the time till the first down-surge is $(\pi/2) \sqrt{A_t \ell / A_p g}$. Time zero is assumed to be that instant at which the water hammer wave reaches the surge tank.

If there is a gradual deceleration of the water column between the pump and surge tank, then the above expressions will not hold, and the correct continuity equation and the rigid column equation for deceleration will have to be solved numerically for successive time intervals.

The fluctuations in tank level may be damped with a throttling orifice. In this case the pressure variations in the line may be more extreme than for the unrestricted orifice. Parmakian (1963) presents charts indicating the maximum upsurge for various throttling losses.

A number of sophisticated variations in the design of surge tanks have been proposed (Rich, 1963). The differential surge tank for

instance includes a small-diameter riser in the middle of the tank.
The tank may have a varying cross section or multiple shafts. Such
variations are more applicable to hydro power plant than pumping
systems, as they are useful for damping the surges in cases of rapid
load variation on turbines.

Discharge Tanks

In situations where the pipeline profile is considerably lower than
the hydraulic grade line it may still be possible to use a tank, but
one which under normal operating conditions is isolated from the
pipeline. The tank water surface would be subjected to atmostpheric
pressure but would be below the hydraulic grade line, as opposed to
that of a surge tank.

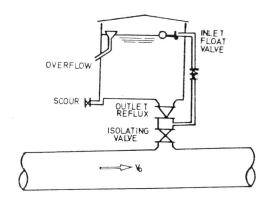

Fig. 4.12   Discharge tank

A discharge tank would normally be situated on the first rise
along the pipeline and possibly on subsequent and successively higher
rises. The tank will be more efficient in reducing pressure variations
the nearer the level in the tank is to the hydraulic grade line. It
should be connected to the pipeline via a reflux valve installed to
discharge from the tank into the pipeline if the pipeline head drops
below the water surface elevation in the tank. Normally the reflux
valve would be held shut by the pressure in the pumping line. A
small-bore bypass to the reflux valve, connected to a float valve in
the tank, should be installed to fill the tank slowly after it has
discharged. Fig. 4.12 depicts a typical discharge tank arrangement.

The use of discharge tanks was reported in detail by Stephenson (1972).

The function of a discharge tank is to fill any low pressure zone caused by pump stoppage, thus preventing water column separation. The water column between the tank and the discharge end of the pipeline (or a subsequent tank) will gradually decelerate under the action of the head difference between the two ends. It may be necessary to prevent reverse motion of the water column which could cause water hammer overpressures – this could be achieved by instal- ling a reflux valve in the line.

A discharge tank will only operate if the water surface is above the lowest level to which the head in the pipeline would otherwise drop following pump stopping. Thus the normal operating head on the first tank along a line should be less than cv_o/g, and subsequent tanks should be successively higher. In cases where the head on the tank is considerably less than cv_o/g, rigid water column theory may be used to calculate the discharge from the tank:

Integrating Equ. 4.1 twice with respect to time, one obtains an expression for the distance the water column travels before stopping. Multiplying this distance by the cross sectional area of the pipeline yields the volume discharged by the tank; $Q = A\ell v_o^2/2gh$, where A is the cross sectional area of the pipe, ℓ is the distance between the discharge tank and the open end of the pipeline, or the next tank if there is one, and h is the head on the tank measured below the hydraulic grade line (or level of the next tank).

Fig. 4.13 depicts the volume discharged from a tank expressed as a fraction of the discharge indicated by the rigid column equation. The coordinates of the graph were obtained from a computer analysis using elastic water hammer theory. It is interesting to note that when the head, h, at the tank is less than approximately 0.5 cv_o/g, the discharge is accurately predicted by rigid column theory, and water hammer overpressures are low. For successively higher heads rigid column theory becomes less applicable and the water hammer overpressures increase. It appears, however, that the discharge from a tank never exceeds that indicated by rigid water column theory, provided the conditions discussed below are complied with.

Fig. 4.13 is only applicable for a single tank along the line. The tank is assumed to be at the pump end of the pipeline, or else there should be an in-line reflux valve immediately upstream (on the pump side) of the tank. Without this reflux valve, a positive wave would return from the tank towards the pumps. In addition to water hammer overpressures which would result at the pumps, the discharge from the tank would exceed that for the assumed case where it only discharged downstream. The maximum overpressures indicated by the chart occur along the pipeline on the discharge side of the tank. The overpessure h' is measured above the discharge head, and is expressed as a fraction of cv_o/g. If the upstream reflux valve is omitted Fig. 4.14 should be used.

For very long pipelines with a number of successively higher peaks, more than one discharge tank may be installed along the line. The tanks should be installed at the peaks where water column separation is likeliest. The lowest head which will occur at any point beyond a tank as the downsurge travels along the line is that of the water surface elevation of the preceding tank.

Fig. 4.15 indicates the discharge from two tanks installed along a pipeline. The chart is also only applicable provided the first tank is located at the pump end of the line, or else a reflux valve should be installed immediately upstream of the tank. The second tank should also have a reflux valve immediately upstream of it.

Fig. 4.15 is really only applicable for cases where the distance between the tanks is equal to the distance between the second tank and the open end of the pipeline; however, the charts hardly vary for different length ratios and Fig. 4.15 could be used as a guide for other cases.

In Fig. 4.15 h_1 is the difference in elevation between the first and second tank, and h_2 is the difference in elevation between the second tank and the head at the discharge end of the pipeline. The corresponding pipe length ℓ_1 is that between the two tanks and ℓ_2 is that between the second tank and the discharge end of the pipe-line. The discharge from the first tank is indicated by broken lines and the discharge from the second tank is indicated by full lines. The relevant h and ℓ are substituted in the value indicated on the

82

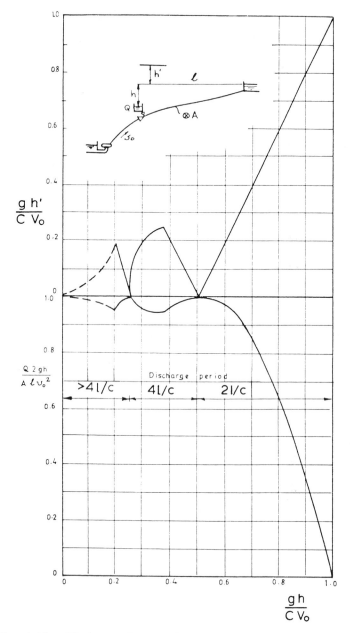

Fig. 4.13   Discharge from tank and maximum head rise with
in-line reflux valve upstream of tank.

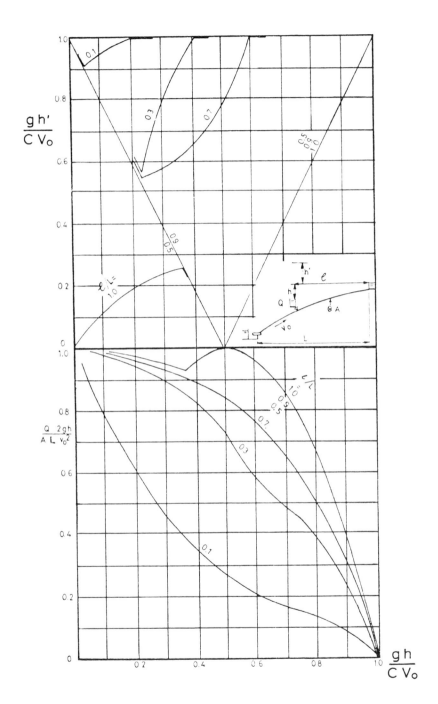

Fig. 4.14   Discharge from tank and maximum head rise
              (no in-line reflux valve).

84

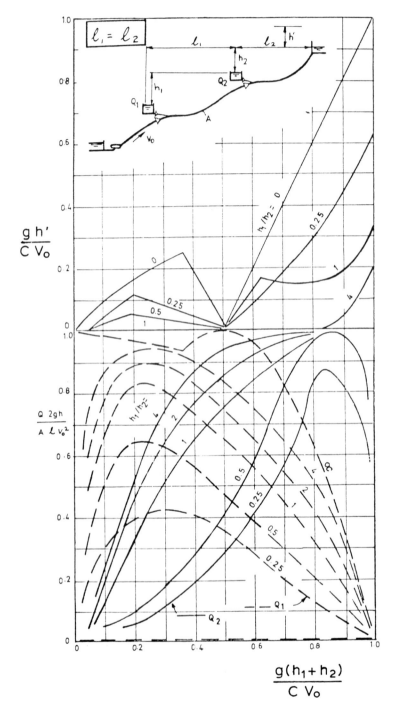

Fig. 4.15 Discharge from two tanks and maximum head rise ($\ell_1 = \ell_2$ and reflux valve upstream of each tank).

vertical scale to obtain the respective discharge Q from each tank.

The maximum overpressures indicated by Fig. 4.15 invariably occur in the downstream side of the second tank. It is, however, possible to limit the lateral extent of the overpressure by judicious location of a further reflux valve between the last tank and the delivery end of the pipeline. In this case the reflux valve on the pump side of the water hammer tank could possibly be dispensed with. The discharge from the tank may then exceed that indicated by the chart, and a water hammer analysis should be carried out based on the theory outlined earlier in the chapter.

The best position for discharge tanks and in-line reflux valves is selected by trial and error and experience. For simple cases the charts presented may suffice, but for complex cases with many peaks or major pipelines with large friction heads, a complete analysis should be carried out, either graphically or by computer. In particular, a final check should be done for flows less than the maximum design capacity of the pipeline.

Even though a number of tanks may be installed along a pipeline, vaporization is always possible along rising sections between the tanks. Provided there are no local peaks, and the line rises fairly steeply between tanks, this limited vaporization should not lead to water hammer overpressures. However, air vessels should be installed on all rising sections. Cases are known where vaporization and vapour bubbles travelling along gently rising mains have resulted in severe cavitation of a line along the top of the pipe.

Air Vessels

If the profile of a pipeline is not high enough to use a surge tank or discharge tank to protect the line, it may be possible to force water into the pipe behind the low-pressure wave by means of compressed air in a vessel (a typical air vessel arrangement is illustrated in Fig. 4.16). The pressure in the vessel will gradually decrease as water is released until the pressure in the vessel equals that in the adjacent line. At this stage the decelerating water column will tend to reverse. However, whereas the outlet of the air vessel should be unrestricted, the inlet should be throttled. A suitable arrangement is to have the water discharge out through a reflux valve, which shuts when the water column reverses. A small-orifice bypass would allow the vessel to refill slowly.

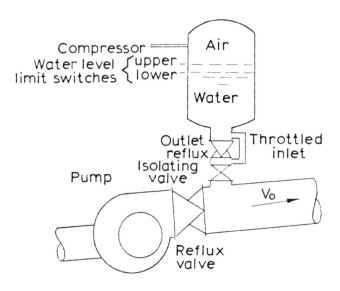

Fig. 4.16   Air vessel

Parmakian (1963) suggests that the air vessel inlet head loss should be 2.5 times the outlet head loss. However, this relationship is empirical and recent research with the aid of computers has enabled the effect of the inlet head loss to be investigated thoroughly. The air vessel design charts reproduced by parmakian were compiled for the ratio of reverse flow head loss to forward flow head loss equal to 2.5. The author has compiled design charts covering an extended range of pipelines and various degrees of inlet throttling (Figs. 4.17 to 4.19). The charts are suitable for checking water column separation along the entire pipeline length, and for reading off the required air volume and degree of inlet throttling corresponding to any specified limit to the overpressures. Outflow throttling is neglected.

The calculations for plotting the charts were performed by computer; it was assumed that the expansion law for the air in the vessel lies between adiabatic ($HS^{1.4}$ = constant) and the isothermal (HS =

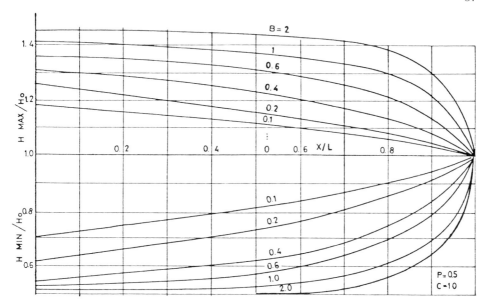

Fig. 4.17   Maximum and minimum pressure envelopes with air vessel.
P = 0.5.

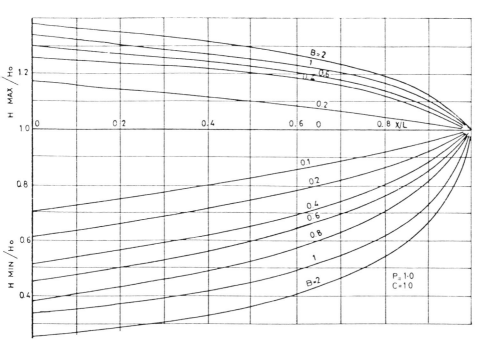

Fig. 4.18   Maximum and minimum pressure envelopes with air
vessel. P = 1.

88

constant). The relationship adopted was $HS^{1.3}$ = constant. It was further assumed that the air vessel was immediately downstream of the pumps, that the pumps would stop instantaneously and that no reverse flow through the pumps was permitted.

The dimensionless parameters associated with the charts are as follows:

Pipeline parameter $\quad\quad\quad\quad P = cv_o/gH_o$

Air vessel parameter $\quad\quad\quad B = v_o^2 AL/gH_o S$

Throttling parameter $\quad\quad\quad C = Z/H_o$

where S is the initial air volume in the vessel, and Z is the head loss through the air vessel inlet corresponding to a line velocity $-v_o$. The outlet head loss was assumed to be negligible. Note that H_o is the absolute head (including atmospheric head) in this case.

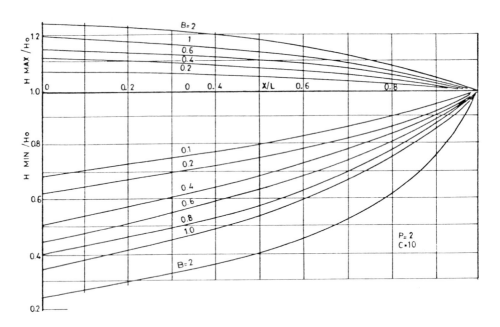

Fig. 4.19   Maximum and minimum pressure envelopes with air vessel. P = 2.

The charts are used as follows to design an air vessel: The pipe-line parameter, P, is calculated from the maximum likely line velocity and pumping head, and the corresponding chart selected from Figs. 4.17 to 4.19. The pipeline profile is plotted on the applicable set of minimum-head curves and a minimum-head envelope selected such that it does not fall below the pipeline profile at any point to cause vaporization. The value of B corresponding to the selected line is used to read off the maximum-head envelope along the pipeline from the same chart. The overpressures actually depend on the degree of inlet throttling represented by C. The normal design procedure is to minimize the overpressures, achieved by setting C approximately equal to 10 (or more). (For a pipeline velocity of approximately 2 m/s and a pumping head of the order of 200 m, the corresponding inlet diameter works out to be 1/10th of the main pipe diameter). If the overpressures indicated thus still cannot be tolerated, it will be necessary to select a smaller value of B (corresponding to a larger volume of air) than indicated by the minimum pressure requirement.

The volume of air, S, is calculated once B is known. The vessel capacity should be sufficient to ensure no air escapes into the pipeline, and should exceed the maximum air volume. This is the volume during minimum pressure conditions and is $S(H_o/H_{min})^{1/1.3}$.

The outlet diameter is usually designed to be about one-half the main pipe diameter. The outlet should be designed with a bellmouth to suppress vortices and air entrainment. The air in the vessel will dissolve in the water to some extent and will have to replenished by means of a compressor.

In-Line Reflux Valves

Normally a reflux valve installed on its own in a pipeline will not reduce water hammer pressures, although it may limit the lateral extent of the shock. In fact, in some situations indiscriminate positioning of reflux valves in a line could be detrimental to water hammer pressures. For instance if a pressure relief valve was installed upstream of the reflux valve the reflux valve would counteract the effect of the other valve. It may also amplify reflections from branch pipes or collapse of vapour pockets.

In-line reflux valves would normally be used in conjunction with surge tanks, discharge tanks or air vessels. Following pump shutdown, the tank or vessel would discharge water into the pipe either side of the reflux valve. This would alleviate the violent pressure drop and convert the phenomenon into a slow motion effect. The reflux valve would then arrest the water column at the time of reversal, which coincides with the point of minimum kinetic energy and maximum potential energy of the water column. There would therefore be little momentum change in the water column when the reflux valve shut and consequently negligible water hammer pressure rise.

There are situations where water column separation and the formation of vapour pockets in the pipeline following pump stoppage would be tolerable, provided the vapour pockets did not collapse resulting in water hammer pressures. Reversal of the water column beyond the vapour pocket could in fact be prevented with an in-line reflux valve at the downstream extremity of the vapour pocket. the water column would be arrested at its point of minimum momentum, so there would be little head rise.

Vaporization would occur at peaks in the pipeline where the water hammer pressure dropped to the vapour pressure of the water. If the first rise along the pipeline was higher than subsequent peaks, the vaporization would be confined to the first peak. The extent of the vapour pocket could be estimated using rigid column theory. The decelerating head on the water column between two peaks or between the last peak and the discharge end would be the difference in elevation plus the intermediate friction head loss. Integrating the rigid water column equation with respect to t, one obtains an expression for the volume of a vapour pocket, $\ell A v_o^2 / 2gh$, where h is the decelerating head, ℓ is the water column length and A is the pipe cross sectional area.

In locating the reflux valve, allowance should be made for some lateral dispersion of the vapour pocket. The valve should be installed at a suitable dip in the pipeline in order to trap the vapour pocket and to ensure proper functioning of the valve doors when the water column returns.

A small-diameter bypass to the reflux valve should be installed to permit slow refilling of the vapour pocket, or else overpressures may occur on restarting the pumps. The diameter of the bypass should be of the order of one-tenth of the pipeline diameter. An air release valve should be installed in the pipeline at the peak to release air which would come out of solution during the period of low pressure.

It is a common practice to install reflux valves immediately downstream of the pumps. Such reflux valves would not prevent water hammer pressures in the pipeline. They merely prevent return flow through the pump and prevent water hammer pressures reaching the pumps.

In some pump installations, automatically closing control valves, instead of reflux valves, are installed on the pump delivery side.

Kinno (1968) studied the effect of controlled closure of a pump delivery valve. Assuming a limited return flow through the pump to be tolerable, he describes how water hammer overpressures can be slightly reduced by controlling the rate of closure of the valve.

Release Valves

There are a number of sophisticated water hammer release valves (often referred to as surge relief valves or surge suppressors) available commercially. These valves have hydraulic actuators which automatically open then gradually close the valve after pump tripping. The valves are normally the needle type, which discharge into a pipe leading to the suction reservoir, or else sleeve valves, mounted in the suction reservoir. The valves must have a gradual throttling effect over the complete range of closure. Needle and sleeve valves are suitably designed to minimize cavitation and corrosion associated with the high discharge velocities which occur during the throttling process.

The valves are usually installed on the delivery side of the pump reflux valves and discharge directly to the suction reservoir. They should not discharge into the suction pipe as they invariably draw air through the throat, and this could reach the pumps.

The valves may be actuated by an electrical fault or by a pressure sensor (as Fig. 4.20).

The valve should open fully before the negative pressure wave returns to the pumps as a positive pressure wave. As the pressure on the top of the piston increases again the valve gradually closes, maintaining the pressure within desired limits. The closing rate may be adjusted by a pilot valve in the hydraulic circuit.

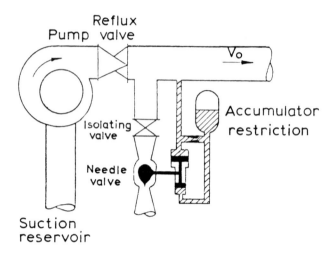

Fig. 4.20   Water hammer release valve arrangement with simplified
            hydraulic actuator

The valve should open fully before the negative pressure wave
returns to the pumps as a positive pressure wave. As the pressure
on the top of the piston increases again the valve gradually closes,
maintaining the pressure within desired limits. The closing rate
may be adjusted by a pilot valve in the hydraulic circuit.

If no overpressure higher than the operating head is tolerable,
the valve would be sized to discharge the full flow at a head equal
to the operating head. Where reliability is of importance, and if
water hammer is likely to be a problem during partial shutdown
of the pumps, two or more release valves may be installed in parallel. They could be set to operate at successively lower delivery
heads.

In the event of normal pump shutdown against throttled delivery
valves, the release valves could be disengaged to prevent their operation.

The types of control valves available as release valves for pumping lines normally cannot open in less than about five seconds. Their
use is therefore limited to pipelines over two kilometers in length.
This method of water hammer protection is normally most economical

for cases when the pumping head greatly exceeds cv_o/g, since the larger the pumping head, the smaller the valve needed.

Since there is no protection against underpressures, a water hammer analysis should be performed to check that water column separation is not a problem.

A less sophisticated valve than the control valves described above, which has been used on small pump installations, is the spring-loaded release valve. The valve is set to open when the pressure reaches a prefixed maximum. Some overpressure is necessary to open the valve and to force water out.

Choice of Protective Device

The best method of water hammer protection for a pumping line will depend on the hydraulic and physical characteristics of the system. The accompanying table summarizes the ranges over which various devices are suitable. The most influential parameter in selecting the method of protection is the pipeline parameter $P = cv_o/gH_o$. When the head cv_o/g is greater than the pumping head H_o, a reflux valve bypassing the pumps may suffice. For successively smaller values of P it becomes necessary to use a surge tank, a discharge tank in combination with an in-line reflux valve, an air vessel, or a release valve. The protective devices listed in the table are arranged in approximate order of increasing cost, thus, to select the most suitable device, one checks down the table until the variables are within the required range.

It may be possible to use two or more protective devices on the same line. This possibility should not be ignored as the most economical arrangement often involves more than one method of protection. In particular the rotational inertia of the pump often has a slight effect in reducing the required capacity of a tank or air vessel. A comprehensive water hammer analysis would be necessary if a series of protection devices in combination is envisaged.

TABLE 4.1

Summary of methods of water hammer protection

Method of Protection (In approximate order of increasing cost)	Required range of variables	Remarks
Inertia of pump	$\dfrac{MN^2}{wALH_o{}^2} > 0.01$	Approximate only
Pump bypass reflux valve	$\dfrac{cv_o}{gH_o} >> 1$	Some water may also be drawn through pump
In-line reflux valve	$\dfrac{cv_o}{gh} > 1$	Normally used in conjunction with some other method of protection. Water column separation possible
Surge tank	h small	Pipeline should be near hydraulic grade line so height of tank is practical
Automatic release valve	$\dfrac{cv_o}{gH_o} << 1$ $\dfrac{2L}{c} > 5$ secs	Pipeline profile should be convex downwards. Water column separation likely.
Discharge tanks	$\dfrac{cv_o}{gh} > 1$	h = pressure head at tank, pipeline profile should be convex upwards
Air vessel	$\dfrac{cv_o}{gH_o} < 1$	Pipeline profile preferably convex downwards

REFERENCES

Kinno, H. and Kennedy, J.F., 1965. Water hammer charts for centrifugal pump systems, Proc. Am. Soc. Civil Engs., 91 (HY3) 247-270.

Kinno, H.,1968 Water hammer control in centrifugal pump systems, Proc. Am. Soc. Civil Engs., 94 (HY3) pp 619-639.

Ludwig, M. and Johnson, S.P., 1950 . Prediction of surge pressures in long oil transmission lines, Proc. Am. Petroleum Inst., N.Y. 30 (5).

Lupton, H.R., 1953 . Graphical analysis of pressure surges in pumping systems, J. Inst. Water Engs., 7.

Parmakian, J., 1963. Water Hammer Analysis, Dover Public. Inc., N.Y.

Rich, G.R., 1963. Hydraulic transients, Dover publics. Inc., N.Y.

Streeter, V.L. and Lai, C., 1963. Water hammer analysis including fluid friction. Proc. Am. Soc. Civil Engs., 88 (HY3) pp 79-112.

Streeter, V.L. and Wylie, E.B., 1967. Hydraulic Transients, McGraw-Hill.

Stephenson, D., 1966. Water hammer charts including fluid friction, Proc. Am. Soc. Civil engs., 92 (HY5) pp 71-94.

Stephenson, D., 1972. Discharge tanks for suppressing water hammer in pumping lines. Proc. Intnl. Conf. on pressure surges,B.H.R.A., Cranfield.

Stephenson, D., 1972. Water hammer protection of pumping lines, trans. S.A. Instn. Civil Engs., 14 (12).

LIST OF SYMBOLS

A – pipe cross sectional area

B – air vessel parameter $v_o^2 AL/(gH_o S)$

C – throttling parameter Z/H_o

d – pipeline diameter

E – modulus of elasticity of pipe wall material

F – pump rated efficiency (expressed as a fraction)

f – Darcy friction factor

g – gravitational acceleration

h – pressure head at an intermediate section of the pipeline

h' – water hammer head rise measured above the delivery head

h_f – friction head loss

h_ℓ – head loss through downstream valve fully open

H – head in pipeline measured above pump suction reservoir level (in case of air vessel design, take H as absolute, i.e. plus atmospheric head)

H_o – pumping head above suction reservoir level

I – pump inertia parameter $MN^2/wALH_o^2$

J – pump parameter $FMN^2 c/180wALv_o gH_o$

K – bulk modulus of water

ℓ – length of an intermediate part of pipeline

L – pipeline length

M – moment of inertia of rotating parts of pump, motor and entrained water (= mass × radius of gyration^2)

N – pump speed in rpm

P – pipeline parameter cv_o/gh_o

Q – volume of water discharged from discharge tank

S – volume of air initially in air vessel

t – time

T – linear valve closure time

v – water velocity in pipeline

v_o – initial water velocity in pipeline

w – weight of water per unit volume

X – distance along pipeline from pump

y – wall thickness of pipe

Z – head loss through air vessel inlet for pipeline velocity $= -v_o$

    – Darcy friction factor

CHAPTER 5

AIR IN PIPELINES

INTRODUCTION

It is recognised that air is present in many water pipelines. The air may be absorbed at free surfaces, or entrained in turbulent flow at the entrance to the line. The air may thus be in solution or in free form in bubbles or pockets. An air pocket implies a relatively large volume of air, likely to accumulate on top of the pipe cross section. The pockets may travel along the line to peaks. There they will either remain in equilibrium, be entrained by the flowing water or be released through air valves.

Air in solution is not likely to present many engineering problems It is only when the pressure reduces sufficiently to permit dissolved air to form bubbles that problems arise. The water bulks and head losses increase. The bubbles may coalesce and rise to the top of the pipe to form large pockets. Flow conditions then become similar to those in partly full drain pipes, except that in a pipeline it is likely that the system, including the free air, will be pressurized.

Air valves are frequently used to eliminate the air which collects on top of the pipe. Air valves of a slightly different design are also used to release large quantities of air during the filling of the line and to draw air in from the atmosphere during vacuum conditions in the line.

PROBLEMS OF AIR ENTRAINMENT

Air drawn in through a pump supply can have a number of effects. Minute air bubbles or air in solution can promote cavitation – the formation of vacuous cavities which subsequently rapidly collapse and erode the pump or pipe. This effect occurs in pump impellers in particular due to the high peripheral speed which lowers pressures. Air drawn in in gulps can cause vibration by causing flow to be unsteady.

98

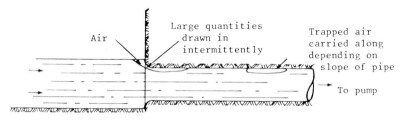

a.   Low level sump

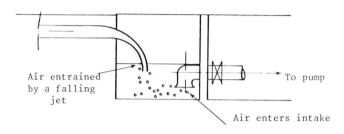

b. Free fall into sump

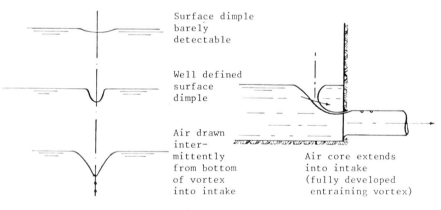

c. Vortex formation

FIG. 5.1   Inlet arrangements conducive to air entrainment.

Air in a rising main, whether in solution or in bubble form, may emerge from solution if ambient pressure or temperature is reduced. Air in free form will collect at the top of the pipeline and then run up to higher points. Here it will either escape through air valves or vent pipes or be washed along by the velocity of the water past the air pocket in the pipe. The latter will result in a head loss due to the acceleration of the water past the air pocket. Even if air is in bubble form, dispersed in the water, the friction head loss along the pipeline will increase. Other problems caused by air present in pipes may include surging, corrosion, reduced pump efficiency, malfunctioning of valves or vibrations.

Air may be present in the form of pockets on the top of the water, in bubbles, micro bubbles or in solution. Bubbles range in size from 1 to 5 mm. Micro bubbles may be smaller, and their rise velocity is so slow (e.g. 1 mm bubble, 90 mm/s, 0.1 mm bubble, 4 mm/s) that they stay in suspension for a considerable time before rising to the surface. In fact turbulence may create an equilibrium concentration profile in the same way that sediment stays in suspension in flowing water. The rise velocity of small air bubbles is

$$w = dg^2/18 \, \nu \qquad\qquad (5.1)$$

where the kinematic viscosity ν of water at 20°C is $1.1 \times 10^{-6} \, m^2/s$, d is bubble diameter and g is gravitational acceleration.

For air bubbles to form readily a nucleus or uneven surface should be present. Then when conditions are correct, air will rapidly be released from solution. The capacity of water to dissolve air depends on the temperature and pressure. At 20°C water may absorb 2% air by volume measured at standard atmospheric pressure. This figure varies from 3.2% at 0°C down to 1.2% at 100°.

AIR INTAKE AT PUMP SUMPS

The major source of air in pumping lines is from the inlet sump or forebay. Here the water exposed to the air will absorb air at a rate depending on its temperature, pressure and degree of saturation of the water.

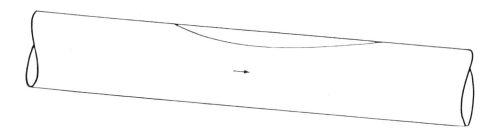

a. Air pocket with subcritical flow past.

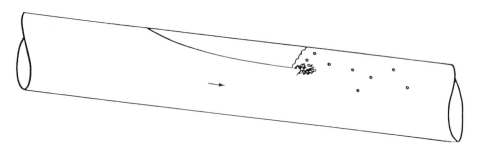

b. Air pocket with super-critical flow past.

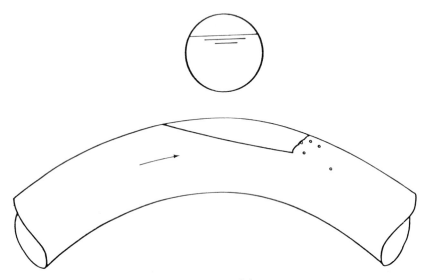

c. Air pocket in equilibrium position.

Fig. 5.2 Air pockets in pipelines.

Air in free form may also be dragged in to the conduit by the flowing water. The formation of a vortex or drawdown outside the suction pipe entrance will entice air into the conduit. The higher the entrance velocity and the greater the turbulence, the more air is likely to be drawn in. This free air may later dissolve wholy or partially when pressures increase beyond the pump. In any case it is carried along the conduit. As pressures again reduce, it may be released from solution.

The configuration of the pump sump has an important bearing on the tendency to draw in air. Extensive studies by Denny and Young (1957), Prosser (1977) and others have indicated how to design sumps to minimize air intake. Circulation in the sump should be avoided by concentric approaches and straight inflow. The inlet should be bellmouthed and preferably facing downwards upstream. Hoods (solid or perforated) above the intake minimize air intake.

The degree of submergence should be as great as possible. Air entraining drops or hydraulic jumps are to be avoided. Fig. 5.1 indicates some inlet arrangements to be avoided.

AIR ABSORPTION AT FREE SURFACES

The rate of diffusion of gas across a liquid interface can be expressed in the form

$$\frac{dM}{dt} = AK \ (C_s - C) \tag{5.2}$$

where M is the mass rate of transfer per unit time t, A is the area exposed, K is a liquid film constant, C is the gas concentration and C_s is the concentration at saturation. K is a function of temperature, viscosity and turbulence. This equation is often written in the form

$$r = \frac{C_s - C_o}{C_s - C_t} = e^{KtA/V} \tag{5.3}$$

where r is the deficit ratio, C_o is the concentration at time 0, and V is the volume of water per surface area A(V/A = depth of water).

HYDRAULIC REMOVAL OF AIR

Air trapped in a pipe and allowed to accumulate, will gradually increase the air pocket size. The water cross sectional area will therefore diminish and the velocity of the water will increase until at some stage, some or all of the air will be dragged along the line by the water. Alternatively a hydraulic jump may form in the pipe entraining air and carrying it away in bubble form, as depicted in Fig. 5.2

The relationship between air pocket volume, washout velocity and pipe diameter has been investigated by a number of workers, and these results are summarized by Wisner et al (1975).

Kalinske and Bliss (1943) produced the following equation for rate of water flow Q at which removal commences:

$$\frac{Q^2}{gD^5} = 0.707 \tan \theta \tag{5.4}$$

where D is the pipe diameter and g is gravitational acceleration and θ is the pipe slope angle. The term on the left hand side of Kalinske's equation is similar to the parameter in the specific momentum and energy diagrams (Figs. 5.3 and 5.4). In fact it points to the possibility that the air is removed if the water depth drops below critical. Then a hydraulic jump forms downstream which could entrain air.

Wisner et al (1975) produced data from which the rise velocity of air pockets may be deduced. They indicate rise velocity is practically independent of slope and is a function of Reynolds number vD/ν and relative volume of air pocket $4U/\pi D^3$ where U is the air pocket volume (see Fig. 5.5)

Hydraulic Jumps

A hydraulic jump draws in air in the form of bubbles. These bubbles exhibit a surprisingly low tendency to coalesce and remain in free form for a long time. Air may be absorbed from the bubbles by the water, but this takes a long time and many of the bubbles rise to the surface before they are absorbed.

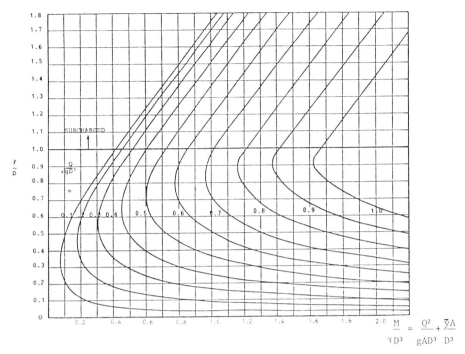

$$\frac{M}{\gamma D^3} = \frac{Q^2}{gAD^3} + \frac{\bar{y}A}{D^3}$$

Fig. 5.3   Specific energy function for circular conduits .

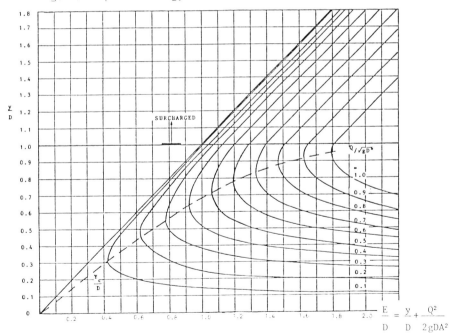

$$\frac{E}{D} = \frac{y}{D} + \frac{Q^2}{2gDA^2}$$

Fig. 5.4   Specific energy function for circular conduits.

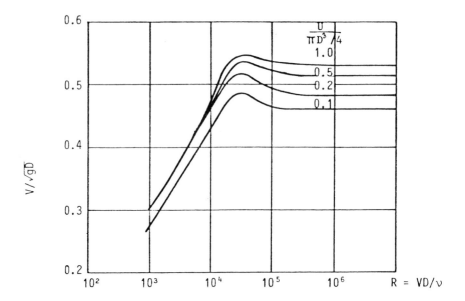

Fig. 5.5 Equilibrium velocity of water for air pockets in a pipe at 10° to 70° to the horizontal.

Kalinske and Robertson (1943) found from model experiments that the rate of air entrainment at a hydraulic jump was given by the equation

$$Q_d/Q = 0.0066(F_1 - 1)^{1.4} \qquad\qquad (5.5)$$

where Q_d is the volumetric rate of air entrainment, Q is the water flow rate and F_1 is the upstream Froude number $v_1/\sqrt{gy_1}$. That relationship was derived for a jump in a rectangular channel with a free surface downstream. Experiments by the author indicate a considerably greater air entrainment rate in closed pipes (see Fig. 5.6)

Free Falls

A jet falling free into a pool of water has a similar air intake effect to a hydraulic jump. Oxygen intake at the base of free falls was studied by Avery and Novak (1978). For an initial oxygen deficiency of 50%, they found aeration at a rate of up to 1.6 kg O_2/kWh for low head losses, decreasing for higher head losses, e.g. at 1 m head loss the aeration efficiency was only 1 kg O_2/kWh. The kW term represents the energy expended or lost in the jump or fall. Assuming 21% oxygen in air, this would indicate aeration rates for air up to 8 kg/kWh.

Water flowing into inlets with a free fall e.g. morning glory type spillways, or even siphon spillways will also entrain a considerable amount of air (Ervine, 1976). Gravity pipelines are therefore as much of a problem as pumping lines.

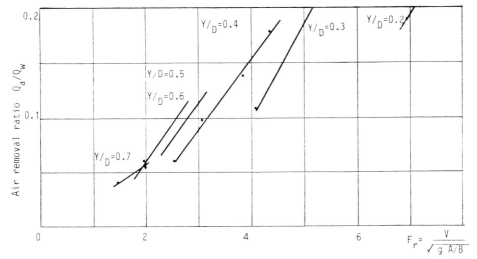

Fig. 5.6   Air removal at hydraulic jumps in circular conduits.

AIR VALVES

Air accumulating in a pocket in a pipe may be released by air valves. These are normally of the 'small orifice' type (typical orifices are up to 3 mm in diameter). The valve opens when sufficient air is accumulated in a chamber to permit a ball to be released from a seat around the orifice.

On the other hand air discharge during filling operations (before

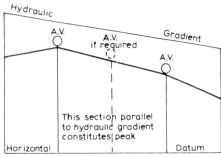

Section of pipeline running parallel to hydraulic gradient and constitutes peak

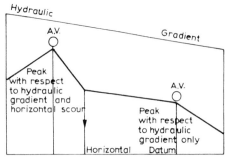

Section of pipeline forming peak with respect to horizontal and also to hydraulic gradient and peak with respect to hydraulic gradient only

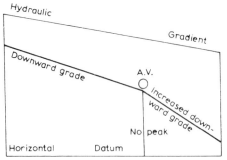

Section of pipeline having downward grade and point of increase of downward grade

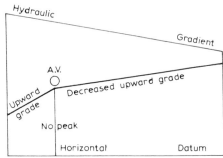

Section of pipeline having upward grade and point of decrease of upward grade

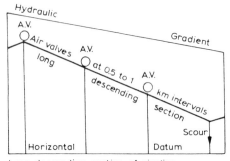

Long descending section of pipeline

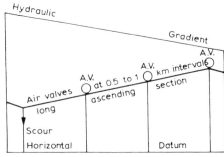

Long ascending section of pipeline

FIG. 5.7   Position of air valves along pipelines

pressurization) and air intake during vacuum conditions in the pipe may be through 'large orifice' air valves, the ball of which is released from the seat around the orifice when pressure inside the pipe is low. Fig. 10.6 illustrates a double air valve with both a small and a large orifice.

The size and spacing of air valves will depend on ambient pressures outside the pipe and permissible pressures inside, the size of pipe and water flow rates.

The equations for the discharge rate through an orifice involve the compressibility of air and are as follows (Marks, 1951). If the pressure beyond the orifice p_2 is greater than $0.53p_1$ (all pressures absolute i.e. gauge plus atmosphere and p_1 is initial pressure), then

$$W \doteqdot Ca \ m_2 \sqrt{\frac{2p_1}{m_1} \ \frac{k}{k-1} \left[1 - (\frac{p_2}{p_1})^{\frac{k-1}{k}} \right]} \qquad (5.6)$$

For $p_2 \leq 0.53p_1$ then the flow becomes critical and flow rate is independent of p_2. Then

$$W = Ca \ m_1 \sqrt{\frac{2p_1}{m_1} \ \frac{k}{k+1} \ (\frac{2}{k+1})^{2/(k-1)}} \qquad (5.7)$$

where W is the mass rate of flow of air, C is a discharge coefficient, a is the orifice area, m is the mass density of air and k is the adiabatic constant.

Normally whether the air valve is for letting air into the pipe or out of it, p_2 is less than $0.53p_1$. Then substituting k = 1.405 for air and C = 0.5 into the last equation, it simplifies to

$$Q_a = 0.34 \ a \ \sqrt{gh/S_a} \qquad (5.8)$$

where Q_a is the volumetric rate of flow of air (in cubic metres per second if SI units are employed) measured at the initial pressure, a is the area of the orifice, h is the initial absolute head in metres of water and S_a is the relative density of air at initial pressure. Since air density is proportional to the absolute head, this simplifies further to

$$Q_a = 100a \qquad (5.9)$$

where Q_a is the air volume flow rate at initial pressure (+ 10 m absolute) in m^3/s and a is in m^2.

Thus to release 1% air from a pipe flowing at a velocity of 1 m/s, and under low head (10 m absolute being the minimum for no vacuum) then (5.8) reduces to

$$0.01 \times 1 \text{ m/s} \times A = 0.34 \text{ a}\sqrt{9.8 \times 10/1.15 \times 10^{-3}}$$

i.e. d = 0.006D   (5.10)

thus the orifice diameter should be about 1% of the pipe diameter to release 1% air by volume in the pipe.

The theory of flow through an orifice was used by Parmakian (1950) to derive equations for the flow through air valves to fill a cavity formed by parting water columns. If initial head (absolute) in metres of water is h_1, h_2 is the head beyond the orifice, g is gravitational acceleration, C is a discharge coefficient for the valve, d is the orifice diameter, D is the pipe diameter, V is the relative velocity of the water column each side of the air valve, k is the adiabatic constant for air and S_a is the relative density of air, then for $h_2 > 0.53 \ h_1$,

$$\frac{d}{D} = \frac{(\Delta V/C)^{1/2}}{\left(\dfrac{2gh_1}{S_{a1}} \dfrac{k}{k-1} \left[1-(\dfrac{h_2}{h_1})^{(k-1)/k}\right]\right)^{1/4}}$$ (5.11)

and for $H_2 \leq 0.53 h_1$

$$\frac{d}{D} = \frac{(\Delta V/C)^{1/2} (h_2/h_1)^{1/2k}}{\left(\dfrac{2gh_1}{S_{a1}} \left[\dfrac{k}{K+1}\right]\left[\dfrac{2}{k+1}\right]^{2/(k-1)}\right)^{1/4}}$$ (5.12)

For air at 300 m above sea level and air temperatures of 24°C plus 20% humidity then S_a is 1.15×10^{-3} and atmospheric pressure is equivalent to h_1 = 9.97 m of water. k is 1.405. For air valves C is about 0.5.

HEAD LOSSES IN PIPELINES

Air suspended in bubble form or in pockets in flowing water will increase the specific volume. The mean velocity is consequently higher to convey a certain volume of water per unit time. The head loss will increase in accordance with an equation such as the Darcy- Weisbach equation

$$h = \frac{\lambda L v^2}{D2g} \qquad (5.13)$$

The velocity v to use will be $v = (1 + f)v_w$ (5.14)

where v_w is the velocity of pure water flowing and f is the volumetric concentration of free air in the pipe.

The head loss around a stationary air pocket is due primarily to a loss of velocity head. If the size of air pocket is such that a hydraulic jump forms, the head loss may be evaluated from Fig. 5.4. If the velocity is subcritical throughout, it may be assumed that the difference in velocity head is lost. It could be established more closely using momentum principles though.

WATER HAMMER

The presence of free air in pipelines can reduce the severity of water hammer considerably. Fox (1977) indicates that the celerity (speed) of an elastic wave with free air is

$$c = \frac{1}{\sqrt{\rho(\frac{1}{K} + \frac{D}{bE} + \frac{f}{p})}} \qquad (5.15)$$

For large air contents this reduces to $c = \sqrt{gh/f}$ (5.16)

where ρ is the mass density of water, K is its bulk modulus, D is the pipe diameter, b its thickness, E its modulus of elasticity, p is the absolute pressure and f is the free gas fraction by volume.

c is reduced remarkably for even relatively low gas contents. Thus 2% of air at a pressure head of 50 m of water reduces the celerity from about 1100 m/s for a typical pipeline to 160 m/s.

The Joukowsky water hammer head is

$$\Delta h = \frac{-c}{g} \Delta v \qquad (5.17)$$

where Δv is the change in velocity of flow. There is thus a large reduction in Δh. If the air collects at the top of the pipe there is no reason to see why the same equation cannot apply. Stephenson (1967) on the other hand derived an equation for the celerity of a bore in a partly full pipe. The celerity derived from momentum principles is for small air proportions

$$c = \sqrt{g \Delta h / f} \qquad (5.18)$$

where Δh is the head rise behind the bore. This indicates a celerity of 158 m/s for $f = 0.02$ and $\Delta h = 50$m.

There is a school of thought which favours the installation of air valves in pipelines as a means of reducing water hammer over-pressures. The intention is primarily to cushion the impact of approaching columns. Calculations will, however, indicate that an excessively large volume of air is required to produce any significant reduction in head. The idea stems from the use of air vessels to alleviate water hammer in pipelines. It will be realised that air in air vessels is under high pressure initially and therefore occupies a relatively small volume. Upon pressure reduction following a pump trip, the air from an air vessel expands according to the equation

$$pU^k = \text{constant} \qquad (5.19)$$

where U is the volume of air. The size of air valves to draw in the necessary volume of air at low (vacuum) pressures will be found on analysis to be excessive for large diameter pipelines.

An unusual problem due possibly to air coming out of solution in a rising main was reported by Glass (1980). Here a thin stream of air along the top of the line supposedly collapsed on a pressure rise after a pump trip. After a number of years the pipe burst along a line along the soffit.

REFERENCES

Avery, S.T. and Novak, P, 1978. Oxygen transfer at hydraulic structures. Proc. ASCE, HY11, 14190, pp 1521-1540.

Denny, D.F. and Young, G.A.J., 1957. The prevention of vortices in intakes. Proc., 7th Con. Int. Ass. Hydr. Res., Lisbon

Ervine, D.A., 1976. The entrainment of air in water. Water Power and Dam Construction. pp 27-30.

Fox, J.A., 1977. Hydraulic Analysis of Unsteady Flow in Pipe Networks. Macmillan, London.

Glass, W.L., 1980. Cavitation of a pump pipeline. Proc. Int. Conf. Pressure surges. BHRA, Canterbury.

Kalinske, A.A. and Bliss, P.H., 1943. Removal of air from pipelines by flowing water, Civil Engineering, ASCE, 13, 10. p 480.

Kalinske, A.A. and Robertson, J.M., 1943. Closed conduit flow, Trans. ASCE. 108, 2205, pp 1453-1516.

Marks, L.S., 1951. Mechanical Engineers Handbook, 5th Ed. McGraw Hill, N.Y. 2235 pp.

Parmakian, J., 1950. Air inlet valves for steel pipelines. Trans., ASCE, 115, 2404., pp 438-444.

Prosser, M.J., 1977. The Hydraulic Design of Pump Sumps and Intakes, BHRA, and CIRIA, London. 48 pp.

Stephenson, 1967. Prevention of vapour pockets collapse in a pumping line. Trans., South African Inst. Civil Engs. 9, (10), pp 255-261.

Wisner, P., Mohsen, F.M. and Kouwen, N., 1975. Removal of air from water lines by hydraulic means. Proc., ASCE. 101, HY2, 11142, pp 243-257.

LIST OF SYMBOLS

a   –   orifice area

A   –   area

b   –   wall thickness

c   –   wave celerity

C   –   discharge coefficient

c   –   gas concentration

C_o –   initial concentration

C_s –   concentration at saturation

d   –   bubble diameter

D   –   pipe internal diameter

E   –   modulus of elasticity

f   –   volumetric concentration of air

F   –   Froude number

g   –   gravitational acceleration

h   –   head

k   –   adiabatic constant

k   –   liquid film constant

K   –   bulk modulus

m   –   mass density of air

M   –   mass rate of transfer or specific momentum

p   –   pressure

Q   –   water discharge rate

Q_d –   volumetric rate of air entrainment

r   –   gas deficit ratio

S   –   relative density or specific gravity

t   –   time

U   –   air pocket volume

v   –   water velocity

V   –   volume of water per surface area

w   –   rise velocity of bubbles

W   –   mass rate of flow of air

y   –   depth of flow

λ –   Darcy friction factor

θ –   pipe slope angle

ν –   kinematic viscosity

ρ –   mass density of water

CHAPTER 6

EXTERNAL LOADS

Low pressure pipes, especially sewers, gravity mains or even large diameter pumping mains should be designed for external loads as well as internal loads. The vertical soil load acting in combination with vacuum pressure inside the pipe could cause the pipe to collapse unless the pipe is adequately supported or stiffened.

SOIL LOADS

The load transmitted to a pipe from the external surroundings depends on a number of factors:

Rigidity of pipe: The more rigid a pipe is relative to the trench sidefill the more load it will take. The sidefill tends to settle, thus causing a large part of the backfill to rest on the pipe. This occurs with flexible pipe too to some extent, as a pipe is supported laterally by the fill and will not yield as much as a free standing pipe.

Type of trench or fill: Fig. 6.1 illustrates various possible installation conditions for pipes. The load transmitted to the pipe varies with the width and depth of trench since friction on the sides of the trench affects the resultant load. Embankment fills may also transmit different loads to a pipe, depending on the relative settlement of sidefill and topfill.

Young and Trott (1984), and Clarke (1968) have developed extensive equations and charts for evaluating soil loads on pipes in various trench and embankment conditions. Although many of their assumptions are subject to question, the fact remains that there is as yet no other theory with which to calculate soil loads on pipes so the engineer must use this theory with discretion (ref. Spangler, 1956).

Trench Conditions

The soil load transmitted to a rigid pipe in a trench depends on the width and depth of trench and the soil backfill properties.

For a normal vertical-sided trench (Fig. 6.1a), fill at the sides of

the pipe will settle more than the pipe, and the sidefill support can
be neglected. On the other hand the friction of the backfill against
the sides of the trench takes some of the load. The cohesion between
trench fill and the sides of the trench is neglected. The friction
developed is therefore proportional to the coefficient of friction
between the fill and the sides of the trench, and the ratio K of the
active horizontal pressure to the vertical pressure in the soil.
Equating the vertical upward forces on any horizontal slice in the
trench to the downward forces, Marston evaluated the total load on a
pipe at depth H (1913):

Upward load per unit length of pipe at depth h + dh below ground
surface:

$$W + dW = W + \gamma B dh - 2K \tan\theta \; W dh / B$$

Solution gives

$$W = C_d \gamma B^2$$

where the load coefficient $C_d = \dfrac{1 - e^{-2K(\tan\theta)H/B}}{2K\tan\theta}$ (6.1)

K = ratio of lateral soil pressure to vertical load.

   = $\dfrac{1 - \sin\phi}{1 + \sin\phi}$ for active soil conditions and cohesionless soil

ϕ = angle of internal friction of backfill

θ = angle of friction between backfill and sides of trench

H = height of fill above pipe

B = trench width

γ = unit weight of backfill material

$K\tan\theta$ normally ranges from 0.11 for soft clays to 0.16 for sands and
coarse crushed stone. C_d is given in Fig. 6.2 for various values for
$K\tan\theta$. Note that for deep trenches, (H/B greater than approximately
10), C_d approaches a limiting value of $1/(2 K\tan\theta)$. This implies that
the side friction in the trench takes more of the load the deeper the
trench. For very shallow or wide trenches (H/B < 1), C_d is
approximately equal to H/B, so

$$W = \gamma HB$$ (6.2)

i.e. most of the backfill load is taken by the pipe. The trench equa-

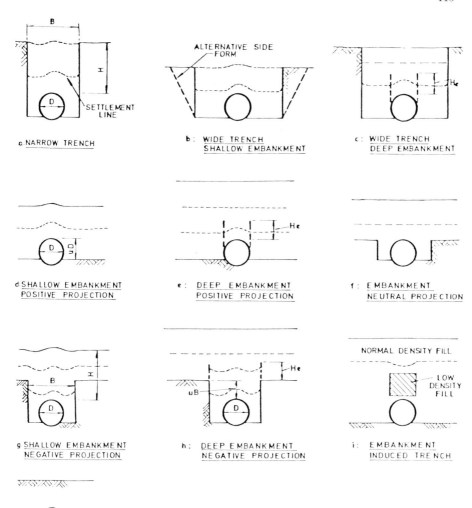

Fig. 6.1   Alternative backfills

tion indicates that the load on a rigid pipe increases indefinitely as the trench width increases. This is not the case and at some trench width the equation no longer applies and embankment conditions (see next section) then apply (see Fig. 6.1b and 6.1c). It is necessary to evaluate the load trying both the trench criterion (using Fig. 6.2) and the embankment criterion (using Fig. 6.3), and to select that load which is least.

In the case of a 'V' trench, the trench width to use is that at the crown of the pipe. In this case ϕ instead of θ in the formula for C_d is used since the shear plane is in the fill not against the sides of the trench. If the side-fill is well compacted and the pipe is relatively flexible then the load on the pipe depends on the pipe diameter and not to the same extent on the trench width:

$$W = C_d \gamma BD \text{ (flexible pipe)} \qquad (6.3)$$

Note that for small H, $C_d = H/B$ and in this case

$$W = \gamma HD \qquad (6.4)$$

Embankment Conditions

A pipe bedded on a firm surface with a wide embankment fill over it is said to be under embankment conditions. If the crown of the pipe projects above the original ground level it is a positive projection (Figs. 6.1d and 6.1e). It is referred to as a complete positive projection if the pipe is rigid and lies completely above natural ground level, and the embankment is relatively shallow. If the pipe crown is below the natural ground level one has a negative projection or trench condition (Figs. 6.1g and 6.1h). For very deep trenches under embankments the conditions may be the same as for the trench condition (referred to here as the complete trench condition). If the crown is level with the natural ground level it is a neutral projection (Fig. 6.1f).

For the complete positive projection case the fill beside the pipe tends to settle more than directly above the pipe. The sidefill tends to drag downwards on the fill directly above the pipe and increase the load on the pipe. The resulting load per unit length of pipe is

$$W = C_c \gamma D^2$$

where the load coefficient

$$C_c = \frac{e^{2K\tan\phi \cdot H/D} - 1}{2K\tan\phi}$$

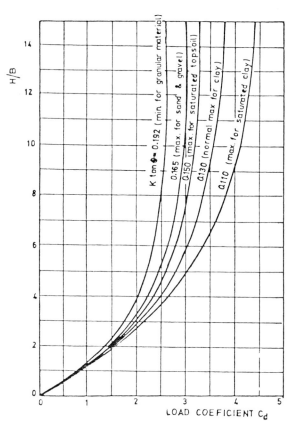

Fig. 6.2   Load coefficient C_d for trench conditions.

The line labelled complete projection condition in Fig. 6.3 gives the value of C_c for $K\tan\phi$ = 0.19, which is the minimum adverse friction condition . For deep embankments, only a certain height of fill above the pipe will settle at a different rate to the sidefill. This height, H_e, depends on the product of the pipe projection ratio u and the settlement ratio s where

$$u = \frac{\text{projection of pipe crown above original ground level}}{\text{diameter of pipe}}$$

s = settlement of adjacent fill originally at crown level—
    settlement of pipe crown
    ——
    reduction in thickness of adjacent fill below crown level

The settlement of adjacent fill includes the settlement of the original ground level and the compaction of the sidefill. The settlement of the pipe crown includes settlement of the bottom of the pipe and vertical deflection of the pipe.

The value of s normally varies from 0.3 for soft yielding foundation soil to 1.0 for rock or unyielding soil foundation. A common value is 0.7 but 0.5 is recommended if the foundation and sidefill are well compacted. The product su is used in Fig. 6.3 to evaluate the resulting load coefficient for various embankment conditions. It is not necessary to evaluate H_e.

The equation for C_c for deep embankments and incomplete positive projection conditions is more complicated than for the shallow embankment condition, but it was evaluated by Spangler for $K\tan\phi$ = 0.19 (i.e. minimum adverse friction) condition, and these values are given in Fig. 6.3.

Fig. 6.3 also gives values of C_c for the case when the top of the pipe settles more than the adjacent backfill (i.e. negative s). Spangler evaluated C_c for $K\tan\phi$ = 0.13 (i.e. minimum favourable friction) for this case, and the values are indicated in Fig. 6.3, labelled complete and incomplete trench conditions since the mechanics are similar to those for trench conditions.

For the negative projection (or incomplete trench) condition (see Fig. 6.1g and 6.1h) with the pipe in a trench under an embankment,

$$W = C_n \gamma B^2 \qquad\qquad\qquad (6.6)$$

where C_n is the same as C_c in Fig. 6.3 for negative su but with D replaced by B in all expressions. In this case the projection ratio is

u = depth of pipe crown below natural ground level
    ——
    width of trench

and the settlement ratio is

s = settlement of natural ground level—settlement of level of top
    fill in trench
    ——
    reduction in thickness of soil in trench above crown

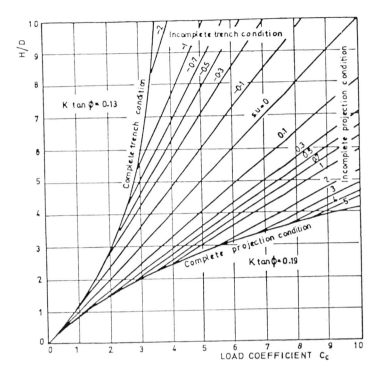

Fig. 6.3   Load coefficient C_c for embankment conditions
(Spangler, 1956).

The settlement of the level of the top of the fill in the trench may include the settlement of the bottom of the pipe, deflection of the pipe and compaction of the fill in the trench above the crown of the pipe. s may range from – 0.1 for u = 0.5, to – 1.0 for u = 2.0.

It is possible to reduce the load on pipes under an embankment by removing a certain amount of compacted fill directly above the pipe and replacing it by lightly compacted fill directly, i.e. inducing a negative projection condition (Fig. 6.1i). For this condition s may range from –0.5 for u = 0.5 to –2.0 for u = 2.0.

The load per unit length on a completely flexible pipe under an embankment whether in positive or negative projection condition is

W   =   γHD   (6.7)

The load on a rigid pipe in a trench up to its crown level (neutral projection condition) is also γHD.

Clarke suggests that the load on a tunnel or thrust bore is similar to that on a pipe in a trench, with a reduction factor allowing for the cohesion of the material above. The assumptions in arriving at this conclusion are not convincing and further theory would be welcome.

Clarke also suggests that uniform surcharges of large extent be treated as embankments and the equivalent depth determined. This is preferable to using pure elastic theory, as some load transfer must take place between soil fills at different densities and soil masses which settle differently.

The soil density which would result in the maximum load on a pipe is the saturated (but not submerged) density. The submerged conditions are normally less severe than for the saturated case, as the water pressure would produce a uniform radial pressure less likely to crack a pipe than a pure vertical load.

Example – Negative projection case

The top of a pipe in a 2 m wide trench is 1 m below natural ground level and there is a 7 m high embankment above the trench. Evaluate the load on the pipe.

Projection ratio $u = \frac{1}{2} = 0.5$

Settlement ratio $s = -0.4$ say.

$H/B = 8/2 = 4$.

From Fig. 6.3, load coefficient $C_n = 3.1$

Load per m of pipe $= C_n \gamma B^2 = 3.1 \times 18000 \times 2^2/1000 = 220 KN/m$

SUPERIMPOSED LOADS

It is customary to use elastic theory to evaluate the pressures transmitted by surface loads to a buried pipe and to assume a semi-infinite, homogeneous, isotropic, elastic material surrounds the pipe. The fact that the sidefill may settle differently to the pipe is ignored in the theory and suitable factors should be applied to decrease the transmitted load if the pipe is flexible. Unfortunately no such factors are available yet.

If the load applied on the surface is of limited lateral extent then the induced pressure on any horizontal plane below the surface decreases with depth. The deeper the pipe, the wider the spread and consequently the lower the maximum pressure. On the other hand the total vertical force must remain equal to the superimposed load.

Formulae for evaluating the transmitted loads due to a surface point load, or a pressure of limited or infinite extent, are given in the sections below. A suitable impact factor should also be applied. The impact factor will vary with the type of load and depth of pipe below the surface, and varies from 1.3 for slow moving vehicles to 2.0 for fast moving vehicles if the pipe is shallow. The impact factor reduces for cover depth greater than the pipe diameter and isolated point loads, but the amount of reduction is debatable and it may be safe to adopt the full value of the impact factor until conclusive figures are available (see also Compstan et al 1978).

Traffic Loads

The following loads are useful for design purposes (see also Concrete Pipe Assn. 1969).

(1)   In fields : two wheel loads each 30 kN spaced 0.9 m apart.

(2)   Under light roads : two wheel loads each 70 kN spaced 0.9m apart.

(3)   Under main roads and heavy traffic (BS 153 type HB load) : eight wheel loads each 112 kN comprising two rows 1.8 m apart of four wheels each 0.9 m apart laterally, measured from inner axle to inner axle.

(4)   Airport runways: four 210 kN wheel loads spaced at 1.67m × 0.66 m.

Pipelines would normally be designed for field loading and strength- ened under heavier traffic. If construction loads are likely to exceed design loads extra topfill should be placed to spread the load during construction (Kennedy, 1971).

Stress Caused by Point Loads

Boussinesque's elastic theory gives the vertical stress at depth H below point load and a distance X horizontally from the point load P as

$$w = \frac{3P \ H^3}{2\pi(H^2 + X^2)^{5/2}} \tag{6.8}$$

The stress at any point due to two or more loads will be the sum of the stresses due to the individual loads. The maximum stress could occur directly under either load or somewhere between them, and the worst case should be selected by plotting the stress between them.

For large pipe diameter, the stress will vary appreciably across the pipe diameter and some average value should be taken. The Concrete Pipe Association (1969) lists the total load in kg/m of pipe for surface wheel loads (1), (2) and (3) in the previous section, and various pipe diameters and depths.

Line Loads

The vertical stress at any depth H below a line load of intensity q per unit length is, by elastic theory,

$$w = \frac{2qH^3}{\pi(X^2 + H^2)^2} \tag{6.9}$$

Uniformly Loaded Areas

The stress intensity at any depth beneath a loaded area on the surface could be evaluated by assuming the distributed load to comprise a number of point loads, and summating the stress due to each load. Fortunately Newmark has performed the integration of infinitesimal point loads to give the vertical stress under the corner of a uniformly loaded rectangular area. Newmark's influence coefficients are given in Table 6.1. To evaluate the vertical stress at any point under a surface loaded area, divide the loaded area into rectangles each with one corner directly above the point in question. Calculate the ratios L/H and Y/H where L and Y are the length and breadth of the rectangle and read the corresponding influence factor from the table. Evaluate the stress due to each rectangle of load by multiplying the load intensity by the influence coefficient for the particular rectangle, and summate the stress due to each such rectangle. If the point in question falls outside the boundaries of the loaded area assume the loaded area extends to above the point and subtract the stress due to the imaginary extensions.

TABLE 6.1   Influence factors for vertical pressure under the corner of a uniformly loaded rectangular area (Capper and Cassie, 1969).

BREADTH / DEPTH $= \dfrac{Y}{H}$	LENGTH OF RECTANGULAR AREA / DEPTH TO STRESSED POINT $= \dfrac{L}{H}$											
	0·1	0·2	0·3	0·4	0·5	0·6	0·8	1·0	1·6	2·0	3·0	4·0
0·1	0·0047	0·0092	0·0132	0·0168	0·0198	0·0222	0·0258	0·0279	0·0306	0·0311	0·0315	0·0316
0·2	0·0092	0·0179	0·0259	0·0328	0·0387	0·0435	0·0504	0·0547	0·0599	0·0610	0·0618	0·0619
0·3	0·0132	0·0259	0·0374	0·0474	0·0559	0·0629	0·0731	0·0794	0·0871	0·0887	0·0898	0·0901
0·4	0·0618	0·0328	0·0474	0·0602	0·0711	0·0801	0·0931	0·1013	0·1114	0·1134	0·1150	0·1153
0·5	0·0198	0·0387	0·0559	0·0711	0·0840	0·0947	0·1103	0·1202	0·1324	0·1350	0·1368	0·1372
0·6	0·0222	0·0435	0·0629	0·0801	0·0947	0·1069	0·1247	0·1361	0·1503	0·1533	0·1555	0·1560
0·8	0·0258	0·0504	0·0731	0·0931	0·1104	0·1247	0·1461	0·1598	0·1774	0·1812	0·1841	0·1847
1·0	0·0279	0·0547	0·0794	0·1013	0·1202	0·1361	0·1598	0·1752	0·1955	0·1999	0·2034	0·2042
1·6	0·0306	0·0599	0·0871	0·1114	0·1324	0·1503	0·1774	0·1955	0·2203	0·2261	0·2309	0·2320
2·0	0·0311	0·0610	0·0887	0·1134	0·1350	0·1533	0·1812	0·1999	0·2261	0·2325	0·2378	0·2391
3·0	0·0315	0·0618	0·0898	0·1150	0·1368	0·1555	0·1841	0·2034	0·2309	0·2378	0·2439	0·2455
4·0	0·0316	0·0619	0·0901	0·1153	0·1372	0·1560	0·1847	0·2042	0·2320	0·2391	0·2455	0·2473

Similar influence coefficients are available for the stresses under loaded circles and strips, but normally any shape can be resolved into rectangles or small point loads without much error. There are also influence charts available for evaluation of stresses under any shape loaded area (Capper and Cassie, 1969).

The stress at any depth under a uniform load of very large extent is the intensity of the surface load since there is no lateral dispersion. This could also be deduced from Table 6.1 for large L and Y.

Effect of Rigid Pavements

A rigid pavement above a pipe has the effect of distributing the load laterally and therefore reducing the stress intensity due to a surface load. The stress at a depth H below a rigid pavement of thickness h, is

$$w = C_r P/R^2 \qquad\qquad (6.10)$$

where $R = 4\sqrt{\dfrac{E\ h^3}{12\ (1 - \nu^2)k}}$ = radius of stiffness

E = modulus of elasticity (28 000 N/mm^2 or 4 000 000 psi for concrete)

h = pavement thickness

ν = Poisson's ratio (0.15 for concrete)

k = modulus of subgrade reaction, which varies from 0.028N/mm^3 for poor support to 0.14 N/mm^3. A typical value for good support is 0.084 N/mm^3

R is normally about three times the slab thickness. C_r is a function of H/R and X/R and is tabulated in Table 6.2 for a single point load P on the surface.

A concrete slab normally has the effect of about five times its thickness of soil fill in attenuating a point load.

TABLE 6.2 Pressure coefficients for a point load on a pavement (Amer. Con. Pipe Assn., 1970)

w Pressure on pipe = $C_r P/R^2$

H Depth of Top of Pipe Below Pavement

P Point Load

R Radius of Stiffness

X Horizontal Distance from Point Load

$\frac{H}{R}$	X/R										
	0.0	0.4	0.8	1.2	1.6	2.0	2.4	2.8	3.2	3.6	4.0
0.0	.113	.105	.089	.068	.048	.032	.020	.011	.006	.002	.000
0.4	.101	.095	.082	.065	.047	.033	.021	.011	.004	.001	.000
0.8	.089	.084	.074	.061	.045	.033	.022	.012	.005	.002	.001
1.2	.076	.072	.065	.054	.043	.032	.022	.014	.008	.005	.003
1.6	.062	.059	.054	.047	.039	.030	.022	.016	.011	.007	.005
2.0	.051	.049	.046	.042	.035	.028	.022	.016	.011	.008	.005
2.4	.043	.041	.039	.036	.030	.026	.021	.016	.011	.008	.006
2.8	.037	.036	.033	.031	.027	.023	.019	.015	.011	.009	.006
3.2	.032	.030	.029	.026	.024	.021	.018	.014	.011	.009	.007
3.6	.027	.026	.025	.023	.021	.019	.016	.014	.011	.009	.007
4.0	.024	.023	.022	.020	.019	.018	.015	.013	.011	.009	.007
4.4	.020	.020	.019	.018	.017	.015	.014	.012	.010	.009	.007
4.8	.018	.017	.017	.016	.015	.013	.012	.011	.009	.008	.007
5.2	.015	.015	.014	.014	.013	.012	.011	.010	.008	.007	.006
5.6	.014	.013	.013	.012	.011	.010	.010	.009	.008	.007	.006
6.0	.012	.012	.011	.011	.010	.009	.009	.008	.007	.007	.006
6.4	.011	.010	.010	.010	.009	.008	.008	.007	.007	.006	.005
6.8	.010	.009	.009	.009	.008	.008	.007	.007	.006	.006	.005
7.2	.009	.008	.008	.008	.008	.007	.007	.006	.006	.006	.005
7.6	.008	.008	.008	.007	.007	.007	.006	.006	.005	.005	.005
8.0	.007	.007	.007	.007	.006	.006	.006	.006	.005	.005	.005

REFERENCES

American Concrete Pipe Assn., 1970. Design Manual-Concrete Pipe, Arlington.

Capper, P.L. and Cassie, W.E., 1969. The Mechanics of Engineering Soils, 5th Ed., Spon., London.

Clarke, N.W.B., 1968. Buried pipelines, MacLaren, London.

Concrete Pipe Assn., 1969. Loads on Buried Concrete Pipes, Tech. Bull. No. 2, Tonbridge.

Compston, D.G., Cray, P., Schofield, A.N. and Shann, C.D., 1978. Design and Construction of Buried Thin Wall Pipes. CIRIA.

Kennedy, H., 1971. External loads and foundations for pipes, J. Am. Water Works Assn. March.

Marston, A. and Anderson, A.O., 1913. Theory of loads on pipes in ditches and tests of cement and clay drain tiles and sewer pipe. Iowa State Univ., Engg. Res. Inst. Bull.

Spangler, M.G., 1956. Stresses in pressure pipelines and protective casing pipes, Proc. Am. Soc. Civil Engrs., 82 (ST5), pp 1054-1115.

Young, D.C. and Trott, J.J., 1984. Buried Rigid Pipes. Elsevier Applied Science, London, pp 230.

LIST OF SYMBOLS

B	–	trench width
C_c	–	load coefficient, embankment condition, positive projection case
C_d	–	load coefficient, trench condition
C_n	–	load coefficient, embankment condition, negative projection case
C_r	–	load coefficient, rigid pavement
D or d	–	diameter
E_s	–	effective modulus of elasticity of soil
E	–	modulus of elasticity
h	–	pavement thickness or variable depth
H	–	depth of cover
H_e	–	depth of fill below plane of equal settlement
I	–	moment of inertia
k	–	modulus of subgrade reaction
k	–	modulus of subgrade reaction
K	–	ratio of lateral to vertical soil pressure
L	–	length of loaded area
M	–	bending moment
N	–	bending or deflection coefficient (subscripts t, b, s, x and y refer to top, bottom and side moments and horizontal and vertical distortions respectively)

p	–	pressure
P	–	point load
q	–	line load per unit length
r	–	radius
R	–	radius of stiffness
s	–	settlement ratio
t	–	wall thickness
u	–	projection ratio
w	–	vertical pressure
w_a	–	lateral soil pressure
w_r	–	permissible ring load
w_v	–	permissible vertical load
W	–	vertical load per unit length
X	–	horizontal distance
Y	–	width of load area
β	–	angle of bottom support
Δ	–	deflection
γ	–	unit weight of soil
θ	–	angle of friction between backfill and sides of trench
φ	–	angle of friction of backfill material
ν	–	Poisson's ratio

CHAPTER 7

CONCRETE PIPES

THE EFFECT OF BEDDING

Non-pressure sewer or drain pipes are designed to withstand external loads, not internal pressures. Various standards specify the design load per unit run of pipe for different classes and the pipes are reinforced accordingly. The main stresses due to vertical external loads are the compressive stress at the haunches and the bending stresses at the crown, the bottom and the haunches. The main stresses are caused by live loads, vertical and horizontal soil loads, self weight and weight of water (internal water pressure and transient pressures are neglected for non-pressure pipes).

Concrete pipes and the other rigid non-pressure pipes are normally designed to withstand a vertical line load while supported on a flat rigid bed. The load per unit length required to fracture the pipe loaded thus is called the laboratory strength. Although the laboratory strength could be calculated theoretically, a number of practical factors influence the theoretical load and experimental determination of the load is more reliable. (The tensile strength of concrete is very uncertain and the effect of lateral constraint of the supports may be appreciable).

Alternative standard testing arrangements for pipes are illustrated in Fig. 7.1. The strength is defined in British Standard 556 as that load which will produce a crack 1/100 inch (0.25 mm) for unreinforced concrete pipes or 80% of the ultimate load for reinforced pipes, or 90% of the 1/10000 inch (0.025 mm) wide crack load for prestressed cylinder type concrete pipes.

The strength using the 3-edge bearing test is usually 5 to 10% in excess of that for the 2-edge bearing test. Recently the 3-edge bearing test was standardized and this is now defined as the laboratory strength.

Pipes laid in trenches are usually supported over a relatively wide length of arc by the bedding. The strength is consequently higher than the laboratory strength as the bending moments are less. Fig. 8.2 summarizes the theoretical bending moment coefficients for pipes supported over various angles (defined in Fig. 8.4)(CPA,1962).

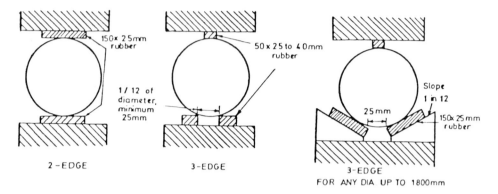

Fig. 7.1   Standard crushing test bearings for rigid pipes

The ratio of field strength to laboratory strength is defined as the bedding factor. Various types of bedding and the corresponding bedding factors are listed below (CPA, 1967; ACPA, 1970):-

TABLE 7.1   Bedding factors

Class A	120° R.C. concrete cradle or arch	3.4
Class A	120° plain concrete cradle or arch	2.6
Class B	Granular bedding	1.9
Class C	Hand shaped trench bottom	1.5
Class D	Hand trimmed flat bottom trench	1.1

For rigid concrete pipes the lateral support of the sidefill in the trench does not add noticeably to the strength.

A factor of safety K of 1.25 to 1.5 is normally used with unreinforced concrete pipe. When pipe is pressurized, it should be designed so that:-

$$\frac{\text{external load}}{\text{field strength}} + \frac{\text{internal pressure}}{\text{bursting pressure}} < \frac{1}{K} \qquad (7.1)$$

Unreinforced or reinforced (not prestressed) concrete pipes are normally only used for non-pressure pipes such as sewers and drains. The concrete may be stressed in tension if the pipe was subjected to internal pressures, and even though concrete has a certain tensile strength, cracks may develop and leaks are likely if the tensile stress is maintained (Kennison, 1950).

Concrete pipes normally do not require lining or wrapping. Special precautions may be necessary for certain liquids, for in-stance sewers are often made with limestone aggregate to maintain the rate of corrosion of the aggregate at the same rate as the cement, thus maintaining an even surface. The friction coefficients of concrete pipes, especially when centrifugally cast, can compare favourably with those of lined steel pipes.

PRESTRESSED CONCRETE PIPES

Prestressed concrete is becoming a popular medium for large-bore pressure pipes. Prestressed concrete competes economically with steel for long pipelines over approximately 800 mm diameter. It has the advantage that the prestressing steel can be stressed to higher stres-ses than for plain-walled pipes. The wall thickness of plain walled steel pipes must be reasonably thick to prevent buckling, collapse and distortion even if the thickness is not required to resist internal pressures. Consequently the use of high-tensile steels is restricted when manufacturing plain-walled steel pipes, but this is not the case with prestressed concrete pipes which are more rigid than steel pipes.

Prestressed concrete pipes are formed by winding pretensioned wires onto a core. The helical winding is subsequently coated. The core forming the barrel of each pipe may be concrete cast vertically in a mould or centrifugally in a horizontal position. The concrete is then steam cured to ensure rapid strength increase before winding. Alternatively the core may be formed with a steel cylinder lined with mortar. The steel cylinder improves watertightness, distributes the concentrated load under the wire, and acts as longitudinal rein-forcing. Plain concrete cores are often reinforced with bars or pre-stressed longitudinally as well as circumferentially. The longitudinal reinforcing resists longitudinal bending in the trench as well as local bending stress in the core walls during circumferential winding. The longitudinal wires are pretensioned and released once the core has set.

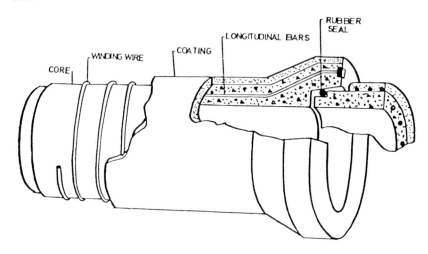

Fig. 7.2   Prestressed concrete pipe

The prestressing wire is wound helically onto the core at a stress close to yield point. It loses a proportion of its stress due to creep and shrinkage before the pipe is put into service. A concrete coating is applied as soon as possible after winding to get as much stress as possible transferred to the coating during the shrinkage of the core.

The tensile strength of the core is neglected in accounting for field pressures but if transient stresses in the field or in the process of manufacture do cause cracking, the cracks soon seal due to the inherent healing properties of concrete. A hydrostatic test is usually done on the pipe immediately after winding but before coating and some seepage of water through the core may be permitted at this stage. Further load tests may be done immediately before dispatch of the pipes from the works, and in the field.

The pipes are usually joined with spigots and sockets sealed with rubber insertions. The socket may be of steel or concrete but care is needed in the design of integral concrete sockets to ensure no stressing or spalling of the socket will occur. Specials (bends and tees) may be fabricated from steel or cast iron, and are fitted with matching spigots and sockets. Small angle bends are normally made up over a number of joints which may take up to 2° deflection, depending on the type of joint.

Circumferential Prestressing

The tensile stress in the prestressing wires immediately after the wire is wound onto the core is less than the applied tensile stress. This is because the core deforms under the prestressing force. At the instant of prestressing any section, the pipe to one side of the section is under stress whereas the other side is free. The strain in the core at the point of winding is half its final strain. The relaxation of the steel stress after winding will therefore correspond to half the total strain of the concrete core and

$$\Delta f_s = \tfrac{1}{2} f_{co} n_1$$

where n_1 is the ratio of the modulus of elasticity of steel to that of the concrete at the time, f_{co} is the core stress after winding and f_s is the steel stress, hence

$$f_{so} = f_{si} - \tfrac{1}{2} f_{co} n_1$$

where f_{si} is the initial steel stress and f_{so} is the steel stress after winding. Since for equilibrium of forces

$$f_{so} S = f_{co} t_c$$

we have
$$f_{co} = \frac{f_{si} S}{t_c + S n_1/2} \tag{7.2}$$

and
$$f_{so} = f_{co} t_c / S \tag{7.3}$$

where S is the steel cross sectional area per unit length of pipe, and t_c is the core thickness.

Circumferential Prestress after Losses

After winding the core, the pipe is usually subjected to a hydro-static test to detect cracks. Any creep which the prestressing steel may undergo has usually taken place by this time, and a certain amount of concrete creep has occurred. The concrete stress just before the test becomes

$$f_{ci} = u f_{co} \tag{7.4}$$

where u is the combined steel and concrete creep coefficient between winding and testing. The corresponding steel stress is:

$$f_{si} = f_{ci} t_c / S \qquad (7.5)$$

The stress for any particular test loading may be calculated using Equs. 7.9 – 7.15 but replacing subscripts 2 by 1 and setting the core thickness t_b equal to zero. The coating should be applied and cured as quickly as possible after testing. This ensures that the shrinkage and creep of the concrete core transfer as much compression to the coating as possible.

The coating adds to the cross sectional area and reduces stresses thereby also limiting the creep of the core.

After the pipe has been cured and is ready for service (usually up to three months after manufacturing) the circumferential tension is taken by the core and the coating:

$$f_{s2} S \doteq f_{c2} t_c + f_{b2} t_b \qquad (7.6)$$

Equating the movement due to elastic compression, creep and shrinkage of the core and coatings to that of the steel:

Deformation of the core:

$$\frac{f_{c2} - f_{c1}}{E_{c2}} + \frac{w_c}{E_{c2}} \frac{f_{ci} + f_{c2}}{2} + v_{c2} = \frac{f_{s1} - f_{s2}}{E_s} \qquad (7.7)$$

(Elastic) (Creep)   (Shrinkage)   (Elastic–steel)

Deformation of coating:

$$\frac{f_{b2}}{E_{b2}} + \frac{w_b}{E_{b2}} \frac{f_{b2}}{2} + v_{b2} = \frac{f_{s1} - f_{s2}}{E_s} \qquad (7.8)$$

(Elastic) (Creep) (Shrinkage) (Elastic deformation of steel)

Now if the properties of the concrete core and coating are similar, the elastic modulus $E_{c2} = E_{b2}$

and the shrinkage coefficient $v_{c2} = v_{b2} = v$ say.

Solving Equs. 7.5 – 7.8 for $f_{s1} - f_{s2}$,

$$f_{s1} - f_{s2} = \frac{\nu E_{c2}(t_c + t_b \cdot \overline{2+w_b}) + t_c f_{c1} w_c}{S(1+w_c) + (t_c + t_b \frac{2+w_c}{2+w_b})/n_2} \tag{7.9}$$

$$\text{(with } \overline{2+w_c} \text{ over } 2+w_b \text{ term and } \frac{w_c}{2}\text{)}$$

Hence from 7.8

$$f_{b2} = (\frac{f_{s1} - f_{s2}}{E_s} - \nu) \frac{2E_{c2}}{2+w_b} \tag{7.10}$$

$$\text{and } f_{c2} = \frac{f_{s2}S - f_{b2}t_b}{t_c} \tag{7.11}$$

Circumferential Stress Under Field Pressure

There are a number of field loading conditions which should be examined including:-

(1)   In open trench with internal test or operating pressure plus self weight and weight of water.

(2)   In backfilled trench with internal pressure plus live load plus self weight and weight of water.

(3)   In backfilled trench with live load plus self weight and pipe empty.

(4)   In backfilled trench with internal pressure, self weight, weight of water and transient pressures.

The most highly stressed sections are usually at the crown (due to prestressing plus bending under external loads plus internal pressures), at the haunches (due to prestressing plus bending plus vertical load plus internal pressure) and at the support (due to prestressing plus bending plus vertical plus horizontal loads plus internal pressure).

Compressive and tensile forces are taken directly on the effective area per unit length of wall comprising the pipe core plus coating plus transformed steel section:

$$A = t_c + t_b + (n_2 - 1) S \tag{7.12}$$

Bending moments due to soil loads, external live loads and weight of pipe plus weight of water are resisted by the effective moment of inertia I per unit length of wall. The distance of the centroid of the effective section from the inside of the core is:

$$e = \frac{(t_c + t_b)^2/2 + (n_2 - 1) \, S \, (t_c + d_s/2)}{t_c + t_b + (n_2 - 1) \, S}$$ (7.13)

where d_s is the diameter of the prestressing wire. The corresponding distance to the outside of the coating is:

$$e_o = t_c + t_b - e$$ (7.14)

The moment of inertia of the effective section is:

$$I = \frac{e^3 + e_o^{\,3}}{3} + (t_c + d_s/2 - e)^2 (n_2 - 1) \, S.$$ (7.15)

Longitudinal Prestressing

Some prestressed pipes are prestressed longitudinally as well as circumferentially. The longitudinal prestressing is added to prevent tensile cracks in the concrete during winding of the core and due to the longitudinal bending moment in service. The longitudinal bars or wires are placed in the mould for the concrete core and pretensioned before casting the core. When the wire is wound onto the core, it causes local bending stress and shear in the core walls. The tension in the longitudinal bars assists in reducing the resulting principle stresses by imparting a compressive stress to the concrete core.

As for the circumferential wires, the longitudinal bars lose some stress due to creep of the steel and concrete. If the bars are initially stressed to f_{siL}, their total cross sectional area is A_s, the pipe core cross sectional area is A_c and the creep relaxation coefficient is u_L, then immediately after release of the longitudinal bars, the stress in the core becomes:

$$f_{coL} = \frac{u_L \, f_{siL} \, A_s}{A_c + n_1 A_s}$$ (7.16)

and the longitudinal steel stress becomes

$$f_{soL} = f_{coL} A_c / A_s$$ (7.17)

The core is normally prestressed circumferentially immediately after release of the longitudinal bars. The helical wire is wound from one end to the other. The radial shear stress in the core, at the point of application of the winding, is derived from elastic theory for pressurized cylinders (Timoshenko and Woinowsky-Krieger, 1959) and is approximately

$$q = 0.54 \, S \, f_{so} / \sqrt{t_c \, D} \tag{7.18}$$

where f_{so} is the circumferential steel stress after winding, and D is the external diameter of the core.

The maximum local longitudinal tensile and compressive stresses due to bending in the core wall during winding have been determined experimentally to be approximately

$$f_{cmL} = 0.3 f_{co} \tag{7.19}$$

where f_{co} is the core circumferential stress after winding.

The effect of the longitudinal steel on the cross sectional area has been neglected in Equs. 7.18 and 7.19 as the steel is near the neutral axis and the expressions are approximate anyway. The maximum tensile stress in the core wall may be derived with the assistance of a Mohr diagram (Fig. 7.3):

If $f_{cmL} > f_{coL}$, which is usually the case, then

$$f_{ctL} = \sqrt{q^2 + \left(\frac{f_{cmL} - f_{coL}}{2}\right)^2} + \frac{f_{cmL} - f_{coL}}{2} \tag{7.20a}$$

and if $f_{coL} > f_{cmL}$

then $$f_{ctL} = \sqrt{q^2 + \left(\frac{f_{coL} - F_{cmL}}{2}\right)^2} - \frac{f_{coL} - f_{cmL}}{2} \tag{7.20b}$$

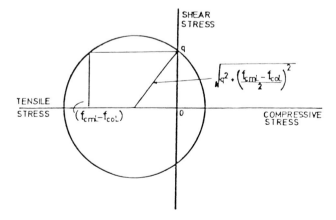

Fig. 7.3   Mohr circle for stress in core during winding.

It will be seen from Equs. 7.20 a and b that the larger the longitudinal prestress, which is proportional to f_{coL}, the smaller is the principle tensile stress in the core f_{ctL}.

Longitudinal Stresses After Losses

The longitudinal bars act to resist longitudinal bending of the pipe in the field. The pipe acts as a beam spanning between uneven points in the bedding. At this stage a certain amount of shrinkage and creep will have occurred in the concrete and some longitudinal compressive stress will have been transferred to the coating. As the longitudinal stress is not high, the concrete creep can usually be neglected. Shrinkage of the core does reduce the tensile stress in the bars and the compressive stress in the core becomes:

$$f_{c2L} = \frac{u_L f_{siL} A_s - v E_s A_s}{A_c + n_2 A_s} \tag{7.21}$$

Once the pipe is in service, longitudinal bending stresses are added to the stress due to prestressing. The extreme fibre bending stresses are

$$f_{c3L} = \frac{MD}{2I} \tag{7.22}$$

where M is the bending moment and the effective moment of inertia of the section is

$$I = \frac{\pi}{64} (D^4 - d^4) + (n_2 - 1) A_s \frac{(d + 2t + d_s)^2}{8} \tag{7.23}$$

Properties of Steel and Concrete

The steel used for winding prestressed concrete pipe should have as high a yield stress as possible and a yield stress of 1650 N/mm² (240 000 psi) is not uncommon. This will ensure that after creep and shrinkage have taken place, the remaining compressive stress in the concrete is fairly high. The steel stress will drop after winding, and it is normal to confine working stresses to less than 50% of yield stress.

Steels with high yield stresses are often brittle and difficult to work with and in fact the ultimate strength may not be much higher than the yield stress. Care is therefore necessary in selecting the prestressing wire. The modulus of elasticity of steel is approximately 200 000 N/mm² (30 × 10⁶ psi).

The steel may creep slightly after prestressing but it is difficult to distinguish between concrete and steel creep between winding and works test, so they are usually considered together. This loss in stress in the steel is typically around 5% and occurs within a few hours after winding.

Concrete cores are cast under vibration or centrifugally and 28 day cube crushing strengths of 60 N/mm² (8 700 psi) are frequently achieved. High early strength (e.g. 50 N/mm² or 7 200 psi) is desirable for winding as early as possible. Compressive stresses up to 50% of the cube strength at the time are permitted during winding (any cracks will most probably heal). Working compressive stresses should be confined to less than 1/3 of the 28 day cube strength.

The bending tensile stress in the core during winding should be less than about 10% of the cube strength at that age, and the calculated bending tensile stress in the core due to longitudinal bending in the field should be limited to about 5% of the cube strength in order to allow for unknowns and to prevent any possibility of cracks developing. No circumferential tensile stress is permitted in the core in the field for normal operating conditions but tension is sometimes permitted under transient conditions. The tensile stress in the coating should be less than about 10% of the cube strength. (This is for buried pipe. Exposed pipes may develop cracks and the tensile stress should be less than this.)

BS 4625 does not permit tensile stress in the core for normal operating plus backfill pressures, but permits a tensile stress of $0.747\sqrt{F}$ if water hammer pressure is included, and a tensile stress of $0.623\sqrt{F}$ during works hydrostatic test. (Where F is the crushing strength of 150 mm cubes at 28 days, all in N/mm².)

The modulus of elasticity of concrete increases with the strength and varies from 20 000 N/mm² (3 000 000 psi) to 40 000 N/mm² (6 000 000 psi) (BSCP 2007, 1970; Ferguson, 1958).

The creep coefficient, w, of concrete is defined by the equation

$$\text{creep strain} = \Delta L/L = w\, f_c/E_c$$

where f_c is the average compressive stress during the time that creep occurs and E_c is the final modulus of elasticity. w varies with time and the method of curing. Creep is high for 'green' concrete and the rate of creep reduces with time. w is approximately 0.3 two days after casting (at the time of factory testing the core) and is 1.3 three months after casting. Hence for the coating, the coefficient 1.3 is used to calculate creep, but for the core, the creep coefficient for the time between factory test and field conditions is 1.3 - 0.3 = 1.0. It may be 30% higher for pipes not properly cured or exposed to the elements in the field.

Shrinkage of concrete also depends largely on the method of curing, typical values of shrinkage per unit length, v, varying from 10^{-4} to 10^{-3}.

Example

Calculate circumferential concrete and steel stresses during winding, after curing and under field conditions for the prestressed concrete pipe described below:

Effective pipe length L = 4 m, bore d_i = 2 000 mm, core thickness t_c = 75 mm, coating t_b = 25 mm, steel winding 5 mm diameter at 15 mm c/c. S = 0.00131 m^2/m of pipe. Longitudinal wires 24 No. × 8 mm diameter, A_s = 0.00121 m^2. Maximum internal pressure 0.8 N/mm^2. Vertical loading due to soil and live load 0.04 N/mm^2, lateral soil pressure 0.5 × vertical = 0.02 N/mm^2. Bottom support effectively over 30° of arc. Neglect selfweight and weight of water inside.

Steel: Longitudinal prestress = 400 N/mm^2. Winding prestress = 1000 N/mm^2. E_s = 200 000 N/mm^2. Creep coefficient u = 0.95.
Concrete: E_{c1} = 30 000 N/mm^2, E_{c2} = 38 000 N/mm^2, Creep w_c = 1.0 from factory test to field test. W_b = 1.3, shrinkage v = 10^{-4}.

n_1 =   200 000/30 000   = 6.7
n_2 =   200 000/38 000   = 5.3

Stresses due to winding

$$f_{co} = \frac{1\ 000 \times 0.0131}{0.075 + 0.00131 \times 6.7/2} = 16.5 \text{ N/mm}^2 \tag{7.2}$$

$$f_{so} = 16.5 \frac{0.075}{0.00131} = 880 \text{ N/mm}^2 \tag{7.3}$$

Works test of core:

$$f_{cl} = 0.95 \times 16.5 = 15.7 \text{ N/mm}^2 \tag{7.4}$$

$$f_{sl} = 0.95 \times 880 = 840 \text{ N/mm}^2$$

after curing:

$$f_{sl} - f_{s2} = \frac{10^{-4} \times 38 \times 10^3 (0.075 + 0.025 \frac{2+1.0}{2+1.3}) + 0.075 \times 15.7 \times 1.0}{0.00131(1 + \frac{1.0}{2}) + (0.075 + 0.025 \frac{2+1.0}{2+1.3})/5.3} \tag{7.9}$$

$$= 76 \text{ N/mm}^2$$

$$f_{s2} = 840 - 76 = 764 \text{ N/mm}^2$$

$$f_{b2} = (\frac{76}{200\ 000} - 10^{-4}) \frac{2 \times 38\ 000}{2+1.3} = 6.4 \text{ N/mm}^2 \tag{7.10}$$

$$f_{c2} = \frac{764 \times 0.00131 - 6.4 \times 0.025}{0.075} = 11.2 \text{ N/mm}^2 \tag{7.11}$$

Under field loading:

Transformed section $A = 0.075 + 0.025 + (5.3-1)0.00131$

$$= 0.105 \text{ m}^2/\text{m} \tag{7.12}$$

Centroid to inner surface $e = \frac{0.1^2/2 + (5.3-1)(0.075+0.005/2)(0.00131)}{0.1 + 0.00131(5.3-1)}$

$$= 0.0515 \tag{7.13}$$

Centroid to outer surface $e_o = 0.1 - 0.0515 = 0.0485$ \tag{7.14}

Moment of inertia $I = \frac{0.0515^3 + 0.0485^3}{3} + (0.075 + 0.005/2 - 0.0515)^2$

$$\times (5.3-1)(0.00131) = 85.9 \times 10^{-6} \text{ m}^4/\text{m} \tag{7.15}$$

Stresses at base B : (tension – ve for concrete stresses)

Tension due to net internal pressure:

$$f_t = \frac{0.8 \times 2 - 0.02 \times 2.2}{2 \times 0.105} = -7.4 \text{ N/mm}^2$$

Bending moment coefficients from Fig. 8.2:

For vertical load, $N_b = 0.235$, and for horizontal load, $N_s = 0.125$

Net bending moment $= 0.235 \times 0.04 \times 2.2/2 - 0.125 \times 0.02 \times 2.2/2$

$$= 0.00755$$

Stress on outer face $f_{b3} = f_{b2} - f_t + \dfrac{Me_o}{I}$

Stress on inner face $f_{c3} = 11.2 - 7.4 - \dfrac{0.00755 \times 0.0515}{85.9 \times 10^{-6}} = -1 \text{N/mm}^2$

Steel stress: $e = 0.075 + 0.005/2 - 0.0515 = 0.026$ (on outer side of centroid)

$f_{s3} = f_{s2} + n_2(f_t - \dfrac{Me}{I})$

$= 764 + 5.3\ (7.4 - \dfrac{0.00755 \times 0.026}{85.9 \times 10^{-6}}) = 791\ \text{N/mm}^2$

Although the stresses at the base are usually the most severe, the stresses at other points on the circumference should be checked similarly, and checks should be done both with and without trans-ient winding, testing and in the field.

REFERENCES

Am. Concrete Pipe Assn., 1970. Design Manual – Concrete Pipe, Arlington.
BSCP 2007 Part 2, 1970. Reinforced and Prestressed Concrete Structures, BSI, London.
Concrete Pipe Assn., 1962. Loads on Buried Concrete Pipes, Tech. Bulletin No.2, Tonbridge.
Concrete Pipe Assn., 1967. Bedding and Jointing of Flexibly Jointed Concrete Pipes, Tech. Bulletin No. 1, Tonbridge.
Ferguson, P.M., 1958. Reinforced Concrete fundamentals, Wiley,N.Y.
Kennison, H.F., 1950. Design of prestressed concrete cylinder pipe, J. Am. Water Works Assn., 42.
Timoshenko, S.P. and Woinowsky-Krieger, S., 1959. Theory of Plates and Shells, 2nd Edn., McGraw Hill, N.Y.

LIST OF SYMBOLS

A – cross sectional area

d – inside diameter

D – outside diameter

e – distance from centre of gravity

E – modulus of elasticity

f – stress (compressive or tensile)

F – crushing strength of 150 mm cubes at 28 days

I – moment of inertia

K – factor of safety

L – length

M – bending moment

n – elastic modular ratio E_s/E_c

q – shear stress

S – steel area per unit length of pipe

t – thickness

u – creep coefficient for steel and concrete before factory test

v – shrinkage coefficient

w – creep coefficient

Subscripts

b – coating

c – core

s – steel

m – bending

q – shear

t – tensile

L – longitudinal

i – initial

o – after winding

1 – at time of factory test

2 – at time of laying

3 – under field pressure

CHAPTER 8

STEEL AND FLEXIBLE PIPE

INTERNAL PRESSURES

The highest pressure a pipe has to resist are normally those due to internal fluid pressure. The pressure is uniform around the pipe and there are no bending stresses except for water and self weight effects. The general equations for the resulting stresses in a hollow cylinder are:

Circumferential wall stress

$$F_w = \frac{p_i d_i^2 - p_o d_o^2}{d_o^2 - d_i^2} - \frac{d_i^2 d_o^2 (p_o - p_i)}{d_o^2 - d_i^2} \frac{1}{d^2} \tag{8.1}$$

Radial stress

$$F_r = \frac{p_i d_i^2 - p_o d_o^2}{d_o^2 - d_i^2} + \frac{d_i^2 d_o^2 (p_o - p_i)}{d_o^2 - d_i^2} \frac{1}{d^2} \tag{8.2}$$

where p is pressure, d is diameter at which the stress is sought, d_i is internal diameter and d_o is external diameter. The equations are for plain stress, i.e. a cylinder free to expand longitudinally.

For the particular case of no external pressure the circumferential stress is a maximum on the inner surface and is

$$F_w max = \frac{d_i^2 + d_o^2}{d_i + d_o} \frac{p}{2t} \tag{8.3}$$

In practice the wall thickness is normally small in comparison with the diameter, and the wall thickness of a steel pipe designed to resist internal pressure is obtained from the formula

$$t = \frac{p(d-t)}{2f \ GJ} \tag{8.4}$$

where t is the wall thickness

p is the internal pressure

d is the external diameter

f is the yield stress of the steel

G is the design factor

J is the joint factor.

Equ. 8.4 overpredicts the true stress by some 1% for every 1% of t/d. On the other hand, just using d instead of (d-t) would under-predict stress by 1% for every 1% t/d.

The design factor G allows a safety margin. What factor to use will depend on the accuracy with which loads and transient pressure have been assessed, the working pressure, economics and the consequences of a burst. If the pipeline is protected against water hammer over-pressure, a factor of 0.6 or even 0.7 is reasonable, but if there are many unknown loads, use of a factor of the order of 0.3 to 0.5 would be wiser.

The joint factor J allows for imperfections in welded seams, and varies from 0.85 for furnace butt-welded joints to 1.0 for seamless pipes.

Small-bore, high pressure pipes are designed to resist internal pressures only, but the larger the diameter and the thinner the walls, the more important become the external loads. Vaporization of the fluid in the pipe may also be possible, in which case this internal pressure should be considered acting in conjunction with external loads.

If the pipe is restrained longitudinally, a longitudinal tension of magnitude vF is induced where v is Poisson's ratio and F is the circumferential wall tension.

TENSION RINGS TO RESIST INTERNAL PRESSURES

Pipes which have to resist very high internal pressures may be strengthened by binding with hoops and spirals. It is often difficult to form pipes in the thick plates which would be required to resist some high pressures. On the other hand it may be possible to form the pipe of thinner plate than would be required for an unstrengthened pipe, and wind it with straps or rods.

The rings required to resist internal pressures do not perform the same function as the stiffening rings required to resist collapse against external loads. The winding to resist internal pressures acts in tension and does not need to be as prominent as the rings to resist external load, which should increase the moment of inertia of the longitudinal section through the pipe and therefore be as high as possible. In fact the tension rings, as they will be called, should be as flat and broad as possible to keep the distance between them to a minimum. There will be high longitudinal bending stresses and circumferential stresses in the pipe wall between rings if the rings are far apart. In comparison, stiffening rings to resist external loads are usually spaced a number of diameters apart. Tension rings may also be used to strengthen old pipes which are to be subjected to higher pressures than originally designed for.

Using the stress–strain relationship and the differential equations of equilibrium for cylinder shells under the action of radial pressure, Timoshenko (1959) derived an equation for the radial displacement of a cylinder subjected to a radial pressure and with tension rings at equi–spacing.

Put $\quad b^4 \quad = \quad \dfrac{12(1-v^2)}{d^2\,t^2}$ (8.5)

and $\quad a \quad = \quad bs$ (8.6)

where v is the Poisson's ratio (0.3 for steel), d is the pipe diameter, t is the wall thickness and s the centre–to–centre ring spacing.

Let $\quad X_1 \quad = \quad \dfrac{\cosh a + \cos a}{\sinh a + \sin a}$ (8.7)

$\quad X_2 \quad = \quad \dfrac{\sinh a - \sin a}{\sinh a + \sin a}$ (8.8)

$\quad X_3 \quad = \quad \dfrac{\cosh a - \cos a}{\sinh a + \sin a}$ (8.9)

$\quad X_4 \quad = \quad \dfrac{\sin (a/2)\,\cosh (a/2) + \cos (a/2)\,\sinh (a/2)}{\sinh a + \sin a}$ (8.10)

Then the ring stress, F_r, is given by

$$F_r \quad = \quad \frac{pd}{2t}\Big/\Big[1 + \frac{bA}{t}\big(X_1 - \frac{X_2^{\,2}}{2X_3}\big)\Big]$$ (8.11)

where A = cross sectional area of the ring.

The circumferential pipe wall stress, F_w, mid–way between rings is given by

$$F_w \quad = \quad \frac{pd}{2t}\Big[1 + 2\big(F_r\frac{2t}{pd} - 1\big)X_4\Big]$$ (8.12)

and the longitudinal bending stress in the pipe under the rings, F_b is given by

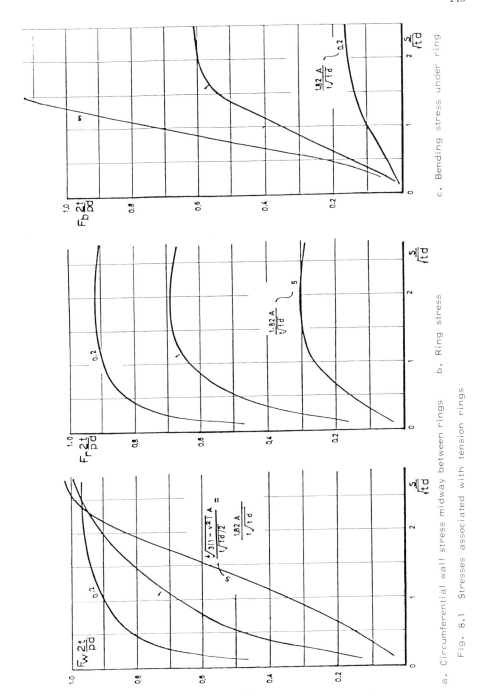

a. Circumferential wall stress midway between rings   b. Ring stress

c. Bending stress under ring

Fig. 8.1 Stresses associated with tension rings

$$F_b = \frac{pd}{2t} \left[\sqrt{\frac{3}{1-v^2}} \left(1 - F_r \frac{2t}{pd}\right) X_2 \right] \tag{8.13}$$

The stresses are indicated by Fig. 8.1, for $v = 0.3$. It may be shown that as s is decreased the ring stress F_r tends to $\frac{1}{2}$ pds/ (ts + A) which could have been anticipated. Note that for small A, F_r tends to equal the circumferential wall stress of a plain pipe, pd/2t. Also, for small s, the circumferential pipe wall stress F_w tends to $\frac{1}{2}$ pds/(ts + A), which would be expected, and the longitudinal beding stress F_b tends to zero.

For large ring spacing (> approx. $2\sqrt{td}$),

$$F_r \text{ tends to } \frac{1}{1 + 0.91A/t\sqrt{td}} \frac{pd}{2t} \tag{8.14}$$

$$F_b \text{ tends to } \frac{1.65}{t\sqrt{td}/A + 0.91} \frac{pd}{2t} \tag{8.15}$$

and F_w tends to $\frac{pd}{2t}$ i.e. that for a pipe without rings.

It may be observed by comparing F_w, F_r and F_b from Figs. 8.1a, 8.1b and 8.1c that the maximum stress for most practical ring sizes is in fact F_w, the circumferential pipe wall stress. Also, F_w is only reduced if $s/\sqrt{td}$ is less than approximately 2.0. In other words, the ring spacing should be less than $2\sqrt{td}$ and the ring cross sectional area A should be of the same order of magnitude as the pipe wall longitudinal cross sectional area between rings, ts, to enable the rings to be of use.

DEFORMATION OF CIRCULAR PIPES UNDER EXTERNAL LOAD

For large diameter and flexible pipes under low internal pressures, the external load is frequently the critical one. Pipes may fail under external load by buckling, overstressing due to arching or bending, excessive deflection.

For elastic rings under plane stress subjected to vertical loads only, Spangler (1956) evaluated the bending moments and deflection at critical points around the circumference. The worst bending moments occur at the crown, the invert or the sides. The bending moments per

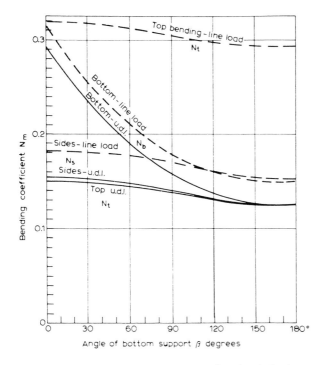

Fig. 8.2 Wall bending coefficients for loaded pipe.

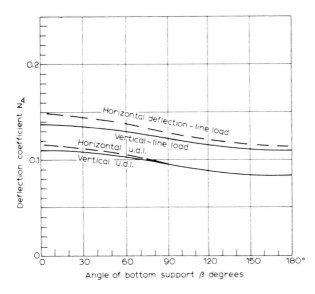

Fig. 8.3 Deflection coefficients for loaded pipe.

unit length are given by an equation of the form

$$M = NWR \tag{8.16}$$

where R is the pipe radius, W is the vertical load per unit length and N is a coefficient. (N_t, N_b and N_s for top, bottom and sides respectively). The vertical and horizontal changes in diameter are practically equal and opposite and are of the form

$$\Delta = N_\Delta WR^3 / EI \tag{8.17}$$

where N_Δ is a coefficient.

The moment of inertia I per unit length of plain pipe wall is

$$I = t^3 / 12 \tag{8.18}$$

Figs. 8.2 and 8.3 give values of N_t, N_b and N_s for the bending moments at the top, bottom and sides respectively, and N_Δ for the vertical deflections of the pipe diameter. The coefficients are for a line load or a load distribution across the width of the pipe (W per unit length), and for different angles of bottom support β , as indicated in Fig. 8.4.

Collapse of a steel pipe will probably not occur until the diameter has been distorted some 10 or 20 percent. In practice deflections up to 5 percent of the diameter are sometimes tolerated. The deflection should normally be limited to about 2 percent to prevent damage to linings, and for pipes with mechanical joints.

The hydraulic properties, i.e. the cross sectional area and wetted perimeter are not affected noticeably for normal distortions.

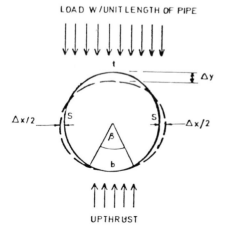

Fig. 8.4   Pipe loading and deflections

Effect of Lateral Support

The lateral support of sidefill in a trench increases the strength of flexible pipes considerably and reduces deformations. Without lateral support to a pipe the ring bending stresses at the soffit and haunches or deflections would limit the vertical external load the pipe could carry. But a pipe in a compacted fill will deflect outwards laterally as it is loaded vertically thereby increasing the pressure of the sidefill against the sides of the pipe. An equilibrium condition may be established with the vertical load being transferred to the haunches by arch action as well as by ring action. The stress due to the arch action is compressive so that the load which the pipe can carry is considerably higher than if the pipe were acting in bending. In the extreme case, the lateral stress will equal the vertical load stress and the pipe wall will be in pure compression, with the stress equal to wd/2t

where w = W/d (8.19)

If the pipe underwent no noticeable lateral distortion the load it could support would be determined by the bending strength plus whatever arch strength is given to the pipe by the soil pressure on the sides of the pipe, i.e. the permissible vertical load per unit area on the pipe is

$$w = w_b + w_a \qquad (8.20)$$

where w_b is the permissible bending load (limited either by ring bending stress or more likely by deflection). w_a is the arching load, equal to lateral soil pressure on the sides of the pipe, which, conservatively, may be taken as the active soil pressure, but is usually greater than this as indicated in subsequent equations. For sand, the active lateral pressure is approximately one third of the vertical pressure due to soil dead load only and for clay it is approximately half the vertical pressure.

If the vertical load is greater than the sum of the ring load plus active lateral soil pressure, the pipe wall will deflect out laterally and increase the lateral pressure. The horizontal stress will decrease away from the pipe and Barnard (1957), using elastic theory, suggests assuming a triangular stress distribution with the horizontal pressure equal to total vertical pressure minus ring load at the pipe wall decreasing linearly to zero at 2.5d away from the pipe wall. The corresponding lateral deflection of each side of the pipe is

$$\Delta X/2 \; = \; 1.25 \; (w-w_b) \; d/E_s \qquad\qquad (8.21)$$

where E_s is the effective modulus of elasticity of the soil. The factor 1.25 should be increased as the lateral deflection increases, since the radial pressure increases as the radius of curvature decreases. The factor becomes 1.4 for a deflection of 2 percent of the diameter and 1.7 for a deflection of 5 percent. The deflection also increases with time, and an additional 25 to 50 percent of the initial deflection can be expected eventually.

If the creep deflection of the soil is large, lateral support can decrease. On the other hand pipe materials which exhibit creep, e.g. plastics, can compensate for this, and could even shed vertical load.

The relationship between stress and strain for the soil should be determined from laboratory triaxial consolidation tests. The effective modulus of elasticity of soil varies widely depending on soil type, degree of compaction or natural density, confining pressure, duration of loading and moisture content. For example it may be as low as 2 N/mm^2 for loose clay or as high as 20 N/mm^2 for dense sands. The modulus is approximately 3 N/mm^2 for loosely compacted fill, 5 N/mm^2 for fill compacted to 90% Proctor density and 7 N/mm^2 for fill to 95% Proctor density. Values higher than 100 have been recorded for moist compacted sands.

The ring load on the pipe is proportional to the elastic deflection as indicated by Equ. 8.17. Taking N_Δ equal to 0.108, which corresponds to bottom support over 30° and the load over the entire pipe width and putting

$$\Delta Y \; = \; \Delta X \; = \; \Delta \text{, then solving for } \Delta/d \text{ from 8.17 and 8.21,}$$

$$\frac{\Delta}{d} \; = \; \frac{0.108wd^3}{8EI+0.043E_s d^3} \qquad\qquad (8.22)$$

For plain pipe $I = t^3/12$, so one has an equation for deflection as a function of diameter, loading, soil modulus and the ratio wall thickness/diameter.

$$\frac{\Delta}{d} \; = \; \frac{0.108w}{0.67E(t/d)^3 + 0.043E_s} \qquad\qquad (8.23)$$

The relationship between deflection and wall thickness for plain pipe is plotted in Fig. 8.5. It will be observed that a steel pipe wall thickness as low as $\frac{1}{2}\%$ of the diameter will be sufficient to restrain distortion to 2% provided the soil modulus is greater than 5MPa and soil load is less than 50 kPa.

Pipes are sometimes strutted internally during backfilling of the trench to increase the vertical diameter and reduce the horizontal diameter. The lateral support increases when the struts are removed and the pipe tends to return to the round shape. The vertical deflection and tendency to buckle are consequently reduced considerably.

STRESS DUE TO CIRCUMFERENTIAL BENDING

It is possible to compute wall stresses due to bending and arching (Stephenson, 1979). If it can be assumed that the load is spread over the full width of the pipe ($\alpha = 180°$) and the bottom support is over 60° ($\beta = 60°$) for flexible pipe, then from (8.17) and Fig. 8.3,

$$\Delta Y = 0.103 w_b d^4 / 8EI \tag{8.17b}$$

Now from (8.21), $\Delta X = 2.5(w - w_b) d / E_s$ (8.21b)

Equating ΔY and ΔX, and solving for w_b the ring load in terms of w the total load,

$$w_b = \frac{wI/d^3}{I/d^3 + 0.006 E_s / E} \tag{8.24}$$

The bending moment in the wall, M, is due to ring load and is a maximum at the base and from (8.16)

$$M = 0.19 w_b d^2 / 2 \text{ hence bending stress } f_b = Mr/I \tag{8.25}$$

where r is the distance from the centre of gravity of the section to the extreme fibre ($t/2$ for plain wall pipe). The balance of the load is taken in arch action of the pipe according to (8.20) so

$$w_a = \frac{0.006 w E_s / E}{I/d^3 + 0.006 E_s / E} \tag{8.26}$$

The bottom wall hoop stress is

$$f_a = w_a d / 2a \tag{8.27}$$

where a is the cross sectional area of wall per unit length (t for plain wall pipe).

The total lateral compressive stress in the wall at the base is f = f_b + f_a, therefore permissible loading per unit area in terms of permissible stress f is

$$w = \frac{(1/d^3 + 0.006E_s/E)f}{0.1\ r/d + 0.003\ dE_s/Ea} \qquad (8.28)$$

Thus the permissible vertical load w is a function of the permissible stress, the relative thickness t/d and the ratio of moduli of soil E_s to steel E. The relationship is plotted in Fig. 8.5 for plain wall pipe, with E = 210 000 N/mm^2, and f = 210 N/mm^2.

For thick pipes, the permissible load increases with wall thickness as more load can be taken in ring bending. For thinner pipes, the deflection becomes large so that the soil side-thrust increases until the pipe is in pure compression, and the limit is due to the wall hoop stress. Actually side wall hoop stress exceeds that at the base, so (8.27) plus (8.25) is not the limiting stress for high arching.

On the same chart is plotted buckling load

$$w_c = \sqrt{32E_s EI/d^3} \qquad (8.29a)$$

The buckling equation, proposed by CIRIA (1978) allows for lateral support. This may be compared with an alternative equation investigated by the Transport and Road Research Laboratory (which is found to overpredict w_c):

$$w_c = (16E_s^2 EI/d^3)^{1/3} \qquad (8.29b)$$

The load w_d giving a deflection of 2% in the diameter is also plotted on the chart from (8.23). The chart thus yields the limiting criterion for any particular wall thickness ratio. The lowest permissible load w is selected in each case from the chart by comparing deflection, overstressing and buckling lines for the relevant t/d and E_s.

Similar charts should be plotted where stiffening rings are used and where alternative material moduli and permissible stresses apply. By careful design it is possible to reach the limit in two or more criteria at the same load. This is called balanced design.

More General Deflection Equations

Spangler (1956) allowed for lateral support to the pipe due to the backfill in a more theoretical way than Barnard. The equation he derived for vertical deflection is:

$$\Delta = \frac{UZWd^3}{8\,EI + 0.06E_s d^3} \tag{8.30}$$

This corresponds to (8.22) if UZ is substituted for N_Δ and a value of 0.15 is assumed for N_Δ in the denominator (instead of 0.108 in 8.22)

Here U = soil consolidation time lag factor (varies from 1.0 to 1.5).

Z = bedding constant (varies from 0.11 for point support to 0.083 for bedding the full width of pipe), normally taken as 0.1.

I = moment of inertia of pipe wall per unit length.

E_s = passive resistance modulus of sidefill.

The pressure inside a pipe may also contribute to its stiffness. Due to the fact that the vertical diameter is compressed to slightly less than the horizontal diameter, the vertical upthrust due to internal pressure becomes greater than the sidethrust by an amount of $2p\Delta$ which tends to return the pipe to a circular shape. A more general expression for vertical deflection thus becomes

$$\Delta = \frac{UZWd^3}{8EI + 0.05E_s d^3 + 2\,UZpd^3} \tag{8.31}$$

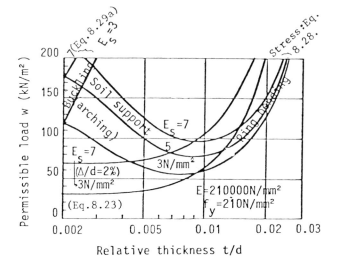

Fig. 8.5   Permissible load on plain pipe

STIFFENING RINGS TO RESIST BUCKLING WITH NO SIDE SUPPORT

Morley (1919) developed a theory for the buckling of stiffened pipes under uniform external pressure. The theory often indicates stiffening ring spacings wider than is considered necessary in practice. The theory also neglects the possibility of failure under combined internal and external pressures and bending. The equations do, however, yield an indication of stiffening ring spacing.

Using an analogy with a strut, Morley developed an equation which indicates the maximum vertical external pressure, w, which a cylindrical shell can take without buckling. The equation allows for no axial expansion, and assumes the wall thickness is small in comparison with the diameter. v is Poisson's ratio,

$$w = \frac{24}{1 - v^2} \frac{EI}{d^3} \qquad (8.32)$$

For plain pipe, $I = t^3/12$, so

$$w = \frac{2E}{1 - v^2} (\frac{t}{d})^3 \qquad (8.33)$$

Experiments indicated a permissible stress 25% less than the theoretical, so for steel,

$$w = 1.65 \, E \, (\frac{t}{d})^3 \qquad (8.34)$$

For thick-walled tubes, the collapse pressure will be that which stresses the wall material to its elastic limit $(w = 2ft/d)$ whereas for intermediate wall thickness, failure will be a combination of buckling and elastic yield. An empirical formula indicating the maximum permissible pressure on a pipe of intermediate thickness is

$$w = \frac{2ft}{d} / (1 - \frac{fd^2}{Et^2}) \qquad (8.35)$$

where f is the yield stress.

The external load may be increased if stiffening rings are used to resist buckling. It was found by experiment that the collapse load, w, is inversely proportional to the distance between stiffening rings, s, if s is less than a certain critical length, L.

From experiment $L = 1.73 \sqrt{\dfrac{d^3}{t}}$ (8.36)

so if $w = 2E \, (t/d)^3$ for stiffening pipe, then for stiffened pipe,

$$w = \frac{L}{s} 2E \, (t/d)^3 = 3.46 \, \frac{Et}{s} \sqrt{\frac{t^3}{d^3}} \qquad (8.37)$$

The actual permissible stress is less than the theoretical due to imperfections in the material and shape of pipe, so the practical ring spacing is given by

$$\frac{s}{d} = \frac{2E}{w} \sqrt[5]{\frac{t^5}{d^5}} \qquad (8.38)$$

If the full elastic strength of the pipe is to be developed to resist vertical external pressures i.e. $w = \dfrac{2ft}{d}$, then the ring spacing should be

$$\frac{s}{d} = \frac{E}{f} \sqrt{\frac{t^3}{d^3}} \qquad (8.39)$$

Rings will only be of use if $w > 1.65E \, (\frac{t}{d})^3$. It should also be ascertained that $w \le \dfrac{2ft}{d}$, which is the elastic yield point.

If $w = 2ft/d$, then rings will only be of use if

$$t/d < f/E = 1/30 \qquad (8.40)$$

for mild steel. Further studies of buckling of stiffened pipe are given by Stephenson (1973) and Jacobsen (1974).

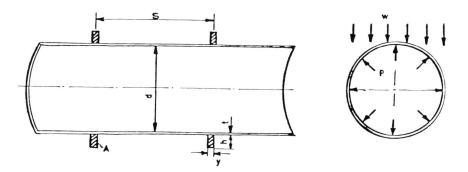

Fig. 8.6   Pipe with stiffening rings

Example 1:

Tension Rings

A 500 mm bore pipe is to take an internal pressure of 5 N/mm^2. The maximum permissible working stress in the wall is to be 100 N/mm^2, and the maximum wall thickness which can be rolled is 10 mm. Assume a 10 mm diameter wire is available for binding the pipe and calculate the required spacing of the binding. An unstiffened pipe could take an internal pressure of

$$\frac{2tf}{d} = \frac{2 \times 10 \times 100}{500} = 4 \text{ N/mm}^2$$

now $\dfrac{pd}{2t} = \dfrac{5 \times 500}{2 \times 100} = 125$, so we require $F/\dfrac{pd}{2t} = \dfrac{100}{125} = 0.8$

Tension ring area parameter $\dfrac{1.82A}{t \sqrt{td}} = \dfrac{1.82 \times \frac{\pi}{4} \times 10^2}{10 \sqrt{10 \times 500}} = 0.2$

By inspection of Figs. 8.1a and 8.1b, the ring spacing required to reduce both the circumferential pipe wall stress and the ring stress to the desired value is given by

$s/\sqrt{td} = 0.45$, $\therefore$ s $= 0.45 \sqrt{10 \times 500} = 32$ mm.

Notice that the amount of steel required is slightly more than if the pipe were merely thickened to take the extra pressure. This effect becomes more noticeable the wider the ring spacing. For wider spacing it is usually the circumferential wall stress which is the limiting stress in the system, so there is little point in using high- tensile rings if the pipe wall is of mild steel.

Example 2:

Stiffening Rings

Design a 3 m diameter steel pipe with stiffening rings to support a uniformly distributed load of 60 kN/m^2 with a soil modulus of 3 N/mm^2.

Maximum steel stress = 103 N/mm^2, delection 2%.

Select t = 10 mm

Then s/d $= (210\ 000/103)\ (.0033)^{3/2}$ (8.39)

$= \quad 0.39 \quad \therefore \quad s = 1177 \text{ mm; Select } 1200 \text{ mm.}$

Use 100 mm high rings with h/b = 10 max for no buckling i.e. ring thickness b = 10 mm

Centroid from centre wall $c = \dfrac{(h+t)bh}{2(bh+ts)} = 4.25 \text{ mm}$ (8.41)

Moment of inertia $I = \dfrac{h^3 b}{12s} + \dfrac{s}{hb}(tc)^2 + \dfrac{t^3}{12} + tc^2$ (8.42)

$= \quad 3104 \text{ mm}^3/\text{mm}$

Extreme fibre distance $r = h + \dfrac{t}{2} - c$

$= \quad 100.8 \text{ mm}$ (8.43)

Deflection: $\Delta/d = \dfrac{0.1 \times 0.060 \times 3^3}{8 \times 210\ 00 \times 3104 \times\ 10^{-9} + 0.05 \times 3 \times 3^3}$ (8.22b)

$= \quad 0.017 \quad = \quad 1.7\%$

Maximum wall stress

$f = \dfrac{(0.1 \times 0.10/3.0 + 0.003 \times 3 \times 3/(210000 \times 0.011))\ 0.06}{3104 \times 10^{-6}/3^3 + 0.006 \times 3/210000}$

$= \quad 1.7 \text{ N/mm}^2 .$

The section could thus be modified. By varying the ring proportions, an optimum or balanced design may be possible i.e. both deflection limit and stress limit are attained simultaneously.

REFERENCES

Barnard, R.E., 1957. Design and deflection control of buried steel pipe supporting earth loads and live loads, Proc. Am. Soc. for Testing Materials, 57.
CIRIA (Construction Industry Research and Information Association), 1978. Design and Construction of Buried Thin-Wall Pipes, Report 78, London, 93 pp.
Jacobsen, S., 1974. Buckling of circular rings and cylindrical tubes under external pressure, Water Power, 26 (12).
Morley, A., 1919. Strength of Materials, Longmans Green & Co., pp 326-333.
Spangler, M.G., 1956. Stresses in pressure pipelines and protective casing pipes. Proc. Am. Soc. Civil Engs., 82(ST5) 1054.
Stephenson, D., 1973. Stiffening rings for pipes, Pipes and Pipelines Intl., 18 (4).
Stephenson, D.,1979. Flexible pipe theory applied to thin-wall nPVC piping, Pipes and Pipeline Intl., 24 (6), 9-17.

Timoshenko, S.P. and Woinowsky-Krieger, S., 1959. Theory of Plates and Shells, 2nd Ed., McGraw Hill, N.Y., Ch. 15.

LIST OF SYMBOLS

a	–	bs, or area per unit length
A	–	cross sectional area of ring
b	–	$\sqrt[4]{\dfrac{12(1 - r^2)}{d^2\, t^2}}$
c	–	distance from centre of wall to centroid of wall section
d	–	pipe diameter
d_i	–	internal pipe diameter
d_o	–	external pipe diameter
E	–	modulus of elasticity
E_s	–	passive resistance modulus of sidefill
f	–	permissible wall stress
F	–	wall stress
F_1	–	bending stress
F_2	–	circumferential stress
F_b	–	bending stress under ring
F_r	–	ring stress, or radial stress in pipe wall
F_w	–	circumferential wall stress
G	–	design factor
h	–	ring height
H	–	depth of backfill above pipe
I	–	moment of inertia
J	–	joint factor
K	–	ratio of active lateral pressure of fill to vertical pressure
L	–	critical ring spacing
M	–	bending moment
N	–	moment or deflection coefficient
P	–	internal pressure
R	–	pipe radius
r	–	distance from centroid to extreme fibre of beam
s	–	ring spacing
t	–	wall thickness

U	–	deflection time lag factor
v	–	Poisson's ratio
w	–	vertical pressure, subscript a refers to arching, b to bending and d to deflection
W	–	vertical load per unit length ($W = wd$)
y	–	ring thickness
z	–	bedding constant
Δ	–	deflection

CHAPTER 9

SECONDARY STRESSES

STRESSES AT BRANCHES

If a hole is cut in the wall of a pipe, the circumferential stress in the wall will increase on each side of the hole. The tangential stress on the horizontal axis at the side of a small elliptical hole in a plate is (Timoshenko and Goodier, 1951),

$$S' = S(1 + 2\ a/b) \tag{9.1}$$

where S is the uniform vertical stress applied to the plate, a is the horizontal axis length and b is the vertical axis length. For a circular hole a = b and the stress S' is consequently three times the applied stress. In the case of a circular hole or branch on a pipe the circumferential stress on each side of the hole or branch may also increase to three times that in the plain pipe. The stress distribution beside a hole in a plate is given in Fig. 9.1. If the pipe wall stress is at all significant it will be necessary to reinforce the wall where a branch pipe occurs. Reinforcing is also desirable to minimize distortion. A common practice especially with low pressure pipework is to use skirts at the branch connection to minimize the entrance head loss. In such cases the size of hole cut in the main pipe is larger than the bore of the branch pipe and reinforcing may be especially necessary. For small branch pipes at right angles to a barrel a suitable method of strengthening the pipe connection is to weld a collar onto the main pipe around the branch pipe (Fig. 9.2). The width of collar may be calculated for any particular thickness by assuming the collar takes the circumferential force which would otherwise have been taken across the diameter of the piece cut out of the main pipe, i.e. 2t'w' = tD, where t' is the collar thickness, w' the width, D the branch pipe outside diameter and t the thickness of the wall of the main pipe. This is assuming that the permissible stress in the collar and main pipe walls are the same, and that the collar width is small compared with the branch pipe diameter.

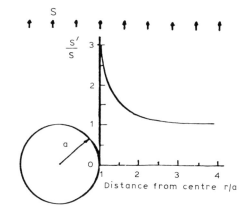

Fig. 9.1   Stress distribution beside a hole in a plate under a uniform stress S.

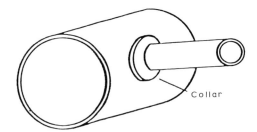

Fig. 9.2   Collar reinforcing at a branch pipe

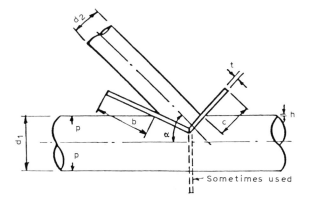

Fig. 9.3   Lateral with external crotch plates

For larger diameter branch pipes and branches at various angles to the main pipe, crotch plates are preferable to collars (Blair, 1946).

Crotch Plates

Various types of stiffening plates have been proposed for reinforcing laterals. The Swiss firm Sulzer Bros. (1941) pioneered the design of crotch plates, and a design aid in the form of a nomograph was later published by Swanson et al (1955). The design was for external plates which would be welded into the crotch at the junction of the pipes. The nomograph was prepared by analysing a large number of cases. Trial sections were selected and revised until forces balanced and deflections were compatible. Only rectangular sections for crotch plates were considered although T sections may have greater rigidity. Figs. 9.3 to 9.7 are design charts for external crotch plates based on Swanson's results, but expressed in dimensionless parameters for use in any system of units. The design charts are for single collar plates (normally for small diameter laterals) and for double crotch plates (for lateral diameter nearly equal to the main pipe diameter).

The method of design using the charts is as follows:

(1) Calculate Pd_1/ft knowing the maximum internal pressure P, the main pipe diameter d_1, the permissible plate stress f and an assumed plate thickness t.

(2) Read off from Fig. 9.5 the crotch plate width expressed in terms of the main pipe diameter d_1. b_1 and c_1 are the crotch plate widths for a 90° lateral with the same diameter as the main pipe.

(3) If the branch is at an angle to the barrel, or the diameters are unequal, read K_b and K_c from Fig. 9.6 and correct the crotch plate width by multiplying b_1 by K_b and c_1 by K_c.

(4) The plate widths are now $b = d_1(b_1/d_1) K_b$ and $c = d_1(c_1/d_1)K_c$.

(5) Read the plate top depth ratio a/d_1 corresponding to b/d_1 or c/d_1 and the angle of the lateral from Fig. 9.7 and hence calculate a, the top depth of each plate.

(6) Check whether either plate width is greater than 30 times the plate thickness. If so, the plate is liable to buckle, so increase the plate thickness and try again. The shape and weight of plates could be optimized by trial.

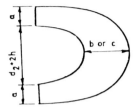

Fig. 9.4 External crotch plates

The shape of the plate should then be drawn for fabrication by developing the line of intersection of the pipes. Single plates may often be bent into shape whereas double plates will be fillet welded together at the intersection point, and if thin enough, may also be bent to follow the intersection of the branch and the barrel.

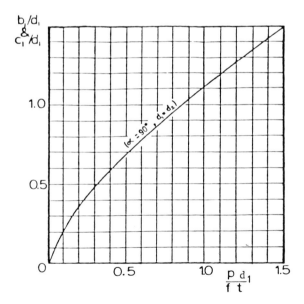

Fig. 9.5 Crotch plate side width

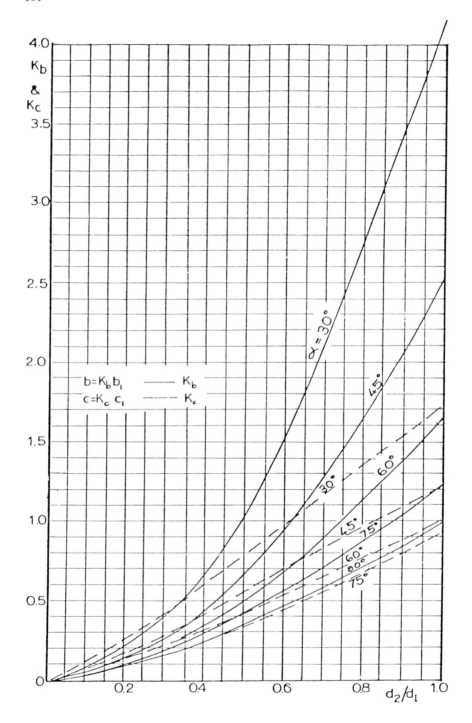

Fig. 9.6   Effect of lateral diameter and angle on crotch plate size.

For large installations a third plate may be required, placed at right angles to the axis of the main pipe (see Fig. 9.3). Although design charts for 3-plate designs are not available, Swanson suggests merely adding a third plate to a 2-plate design. Discretion should be used in reducing the thicknesses or widths of the two crotch plates in such circumstances. The deflections are reduced considerably by adding a third plate though.

Whether crotch plates should be used at all is also a matter of judgement. The author advises using crotch plates if the internal design pressure P is greater than $yt_1 \alpha °/135Gd_2$ where G is the factor of safety, t_1 is the wall thickness of the main pipe, d_2 is the diameter of the branch pipe and α is the angle between the axis of the branch pipe and barrel. y is yield stress.

Internal Bracing (Stephenson, 1971)

Certain disadvantages in external crotch plates prompted Escher Wyss Ltd. (Suss and Hassan, 1957) to investigate internal bracing for pipe 'Y's. External crotch plates act in bending, consequently use material inefficiently and require more steel than internal crotch plates which could be in pure tension.

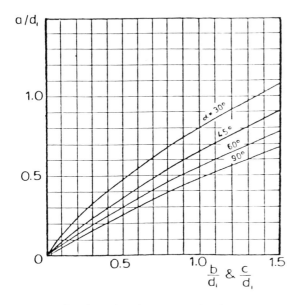

Fig. 9.7 Crotch plate top depth

An internal web may also act as a guide vane. This effect, combined with a gradual taper of the pipes at the junction, reduces head losses for some flow configurations. For the case of water flowing through two pipes to a confluence, the head loss with internal webs is considerably less than with a standard 'T' junction. The arrangement of the branch pipe joining the main pipe at a skew angle with both pipes gradually flared at the confluence, together with internal bracing for the acute angle (see Fig. 9.8) is recommended for pumping station delivery pipes. For water flowing from a main pipe into two branch pipes, such as pumping station suction pipes and penstocks for hydro-electric stations, hydraulic model tests should be performed to determine the best angles of divergence and taper, and the arrangement of the internal bracing.

The overall weight of steel for internal bracing is less than for external crotch plates. The compact arrangement facilitates transport and reduces excavation costs.

Unless the pipes are designed with cone-shaped flares at the confluence, internal bracing will protrude into the flow paths, causing increased velocities and higher head losses than with external crotch plates only. On the other hand if the expansion is too abrupt there will be a high head loss. The head loss coefficient for conical diffusers increases rapidly with the angle of flare once the angle between the wall and axis exceeds about $7\frac{1}{2}$ degrees (Rouse, 1961). A suitable compromise for the angles of flare for a 45-degree confluence appears to be $7\frac{1}{2}$ degrees to the axis for the main pipe and 15 degrees to the axis for the branch pipe.

The use of internal bracing confines the angle between the branch and main pipes to a practical range. For confluence angles approaching 90 degrees ('T' junctions) the bracing web would obstruct the flow in the main pipe. For very small angles of confluence the hole cut in the main pipe becomes larger and the bracing would be impractically heavy. The design charts presented here are confined to a 45-degree confluence, and for any other angle of approach the branch pipe would have to be constructed with a bend immediately before the flare. Note that the use of internal bracing is confined to the acute angle, as internal bracing at the obtuse angle would obstruct flow. The obtuse angle must be strengthened with external crotch plates.

To minimize head losses, the diameters of the main incoming pipe, the branch pipe and the confluent pipe should be such that the velocities in all pipes are approximately equal.

If all branches of the confluence are tangential to a sphere whose centre lies at the intersection of their axes, the intersection of the surfaces will be in planes. This means that the bracing webs are flat. There may be a kink in the obtuse angle intersection line where the main pipe flare meets the confluent pipe (see Fig. 9.8). To eliminate the resulting problems of torsion, plate 'B' may be continued beyond the point where it should bend. The pipe walls would be strong enough to transfer the loads to this plate, and the resulting arrangement simplifies fabrication.

To determine the profiles of the lines of intersection of the cone surfaces, equations were set up relating the cone radii at any point and the horizontal distance along the intersecting plane to the distance from the intersection of the axes along the main axis. Using the fact that the elevations of the surface of both cones are equal on the intersection line, the coordinates of the intersection line were derived. The angle of the intersection plane to the main axis, β, is given by the equation $\tan \beta = (\cos \phi - \cos \alpha \cos \theta)/\sin \alpha \cos \theta$, where α is the confluence angle, ϕ is the angle of the branch cone surface to its axis and θ is the angle of the main cone surface to its axis.

The intersection profiles were obtained by computer analysis proceeding in steps along the intersecting plane. This was done simultaneously with the stress analysis.

Although it is possible to study the stresses in the bracing for simple pipe 'Y's analytically, the general case is very complex and must be solved by computer, proceeding in small increments along the intersection line. In fact previous studies have relied on model studies to obtain the stresses at the intersecting plane.

The analysis was based on membrane theory i.e. the pipe walls can take no bending stress. The circumferential wall stress in a cone is $PR/(\cos \theta)$ per unit width and the longitudinal wall stress $PR/(2$

cos θ) per unit width where P is the liquid pressure and R is the radius of the cone perpendicular to the axis.

The forces due to the cone wall stresses on an element of the intersection line were resolved into three components, perpendicular to the intersection plane, vertically and horizontally in the plane.

The results indicated that the perpendicular forces on the inter-section plane were in all cases zero. The position and direction of the resultant force in the plane, to the side of four points along the intersection line, was plotted graphically. It was observed that the line of action of the resultant was constantly inside the intersection plane and acting in tension on the internal bracing plate.

The most economical internal bracing plate would be one in pure tension. This state could be assured if the resultant force fell on the centre line of the plate. Thus the width of the widest internal bracing plates indicated in Figs. 9.8 to 9.10 at the centre, or crown, is twice the distance from the crown to the resultant force. Hence the thickness of the plate may be derived in terms of the permissible plate stress, and the width anywhere else calculated knowing the plate thickness and position and magnitude of the resul-tant force on the intersecting line of the cones. In some cases it will be found that the most economical bracing plate thickness is impractically thin, in which case a thicker plate may be selected, and the corresponding shape interpolated from the charts. The nar-rower plates are in combined bending and tension and consequently contain more steel than the plate in pure tension. The width of the plate perpendicular to the resultant force at various points was obtained from the equation for extreme fibre stress, $f = 6F \ (u-w/2) /tw^2 + F/tw$, where f is permissible plate tensile stress, F is magni-tude of resultant force which acts at a distance u from the point in question on the intersection line, w is plate width and t is plate thickness. A similar analysis was performed for the external crotch plates bracing the obtuse angle.

The charts, Figs. 9.8 to 9.10, were compiled and facilitate the design of crotch plates for a 45 degree 'Y' and various angles of flare to the main pipe and branch pipe. The most suitable flare angles should be selected from hydraulic considerations. Model tests to determine head losses could be conducted if warranted.

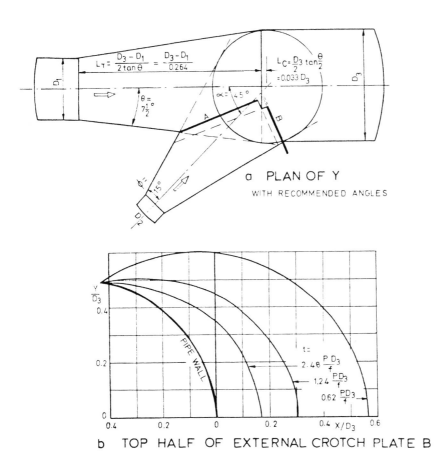

a PLAN OF Y

WITH RECOMMENDED ANGLES

b TOP HALF OF EXTERNAL CROTCH PLATE B

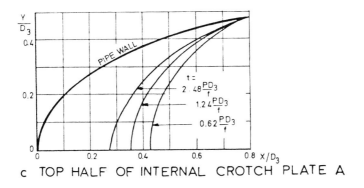

c TOP HALF OF INTERNAL CROTCH PLATE A

Fig. 9.8   Crotch plates for $7\frac{1}{2}°/15°$ confluence.

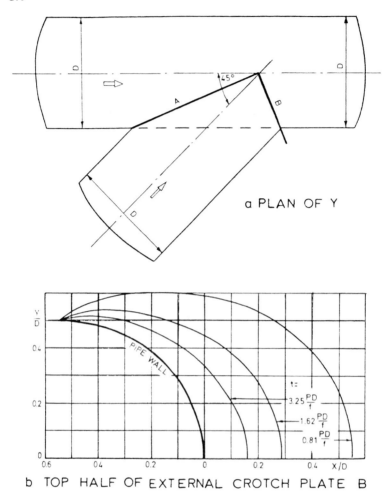

a PLAN OF Y

b TOP HALF OF EXTERNAL CROTCH PLATE B

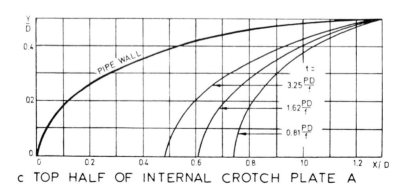

c TOP HALF OF INTERNAL CROTCH PLATE A

Fig. 9.9   Crotch plates for plain pipe confluence

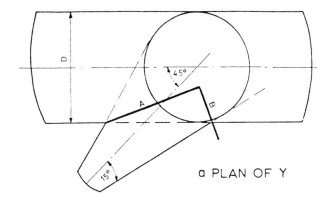

a PLAN OF Y

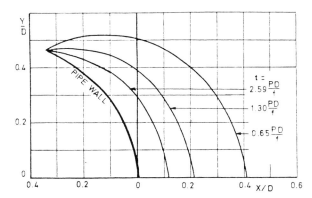

b TOP HALF OF EXTERNAL CROTCH PLATE B

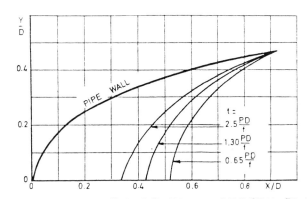

c TOP HALF OF INTERNAL CROTCH PLATE A

Fig. 9.10 Crotch plates for plain pipe with 15° taper confluence pipe.

The design procedure for the crotch plates would then be to select a trial plate width, say the maximum indicated, and compute the plate thickness, t, from the corresponding formula. To ensure that the crotch plate for the obtuse angle is rigid enough to withstand buckling under bending compression, the plate thickness should not be less than w/30. The plate thicknesses should also preferably not be less than the wall thickness of the main pipe, to ensure durability against wear. After selecting the plate thicknesses, the corresponding plate shapes are interpolated from the applicable chart.

STRESSES AT BENDS

The wall stresses in a fabricated or cast iron bend are higher than those in a plain pipe with the same wall thickness. Longitudinal stresses are induced as the bend tends to straighten out, local bending stresses are caused by anchor blocks thrusting against the outside wall, and circumferential stresses on the inside of the bend are magnified because the length of wall on the inside is less than on the outside of the bend. The increase in circumferential stress is normally more severe than the increase in longitudinal stress and it is safe to design only for the worst circumferential stress on the inside of the bend. The wall thickness on the outside of the bend could actually be reduced if account is taken only of circumferential stress, but it is safe to keep the same thickness as on the inside of the block. Circumferential stresses are calculated below.

Using membrane theory i.e. assuming the pipe wall takes no bending stress, the circumferential force on a length of wall on the inside of the bend is equated to the thrust due to the internal pressure on the inside segment of pipe shown shaded in Fig. 9.11,

$$ft \ (R{-}D/2) \ \theta = P(R{-}D/4) \ \theta D/2$$

$$f \ = \ \frac{PD}{2t} \ \frac{(R{-}D/4)}{(R{-}D/2)} \tag{9.2}$$

i.e. the normal wall thickness at the bend should be increased in the ratio $(R{-}D/4) \ / \ (R{-}D/2)$ where R is the radius of the bend and D is the pipe diameter.

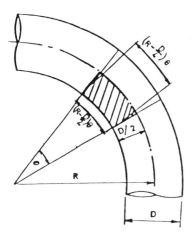

Fig. 9.11   Stress concentration at a bend

In addition to the so-called secondary stress at bends, junctions etc., discussed above, there are sometimes tertiary stresses in pipework caused by relative movements, elastic strains or temperature variations in piping systems. These stresses are real and should be designed for in exposed chemical and plant pipework or at large pipeline interconnections if no movement joints or thrust anchors are installed. A pipe system can be treated as a structural system with axial and lateral thrusts and movements. The system could be analysed using structural design techniques such as moment-distribution, slope-deflection or finite element methods, many of which are available in the form of standard computer programs. (See also Crocker and King, 1967).

THE PIPE AS A BEAM

Longitudinal Bending

Pipes should normally be designed to resist some bending in the longitudinal direction even if they are to be buried. Unevenness or settlement of the bedding could cause sections to cantilever or span between supports. Three possible bending patterns together with the corresponding critical bending moments are indicated in Fig. 9.12.

The maximum span which a simply supported pipe could accommodate is calculated below. The maximum fibre stress is

$$F_b = M/Z \tag{9.3}$$

where M is the bending moment and Z is the section modulus. For a pipe whose wall thickness, t, is small in comparison with the internal diameter D,

$$Z = \pi D^2 \, t/4 \tag{9.4}$$

$$\text{So if } M = WL^2/8 \tag{9.5}$$

$$F_b = WL^2/(2\pi D^2 \, t) \tag{9.6}$$

$$\text{or } L^2 = 2\pi D^2 \, t \, F_b/W$$

and if a pipe of specific weight γ_s is conveying water at a specific weight γw

$$W = \gamma_w \, \pi D^2/4 + \gamma_s \, \pi D t \ \text{N/m}$$

$$L = \sqrt{8DtF_b/(\gamma_w D + 4\gamma_s t)} \tag{9.7}$$

Thus for D = 1 m, t = 0.012 m, F_b = 100 N/mm^2, γ_w = 10 000 N/m^3, γ_s = 80 000 N/m^3 then the maximum permissible simply supported span L = 26 m or 86 ft.

For low internal water pressures, buckling of the walls may also be a problem and the possibility should be investigated (especially at the supports where the shear stress is high). See also Pearson, 1977.

Pipe Stress at Saddles

The maximum local stress in a continuous or simply supported pipe supported in a saddle is (Roark, 1954):

$$F = (0.02 - 0.00012 \, (\beta - 90°))(Q/t^2)\text{Log}_e \, (\tfrac{r}{t}) \tag{9.8}$$

where Q is the total saddle reaction. Note that for a simply supported span, Q is the reaction due to two ends resting on the support, i.e. due to the weight of pipe of length equal to the span, as for a continuous pipe. If Q is in Newtons and t in mm then F is in N/mm^2.

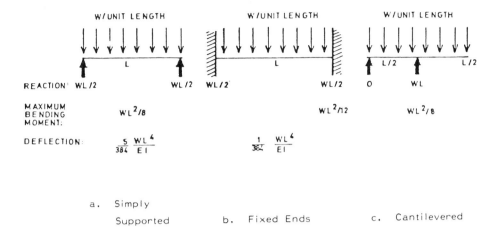

Fig. 9.12   Beam moments and deflections

The angle of bottom support, β , is normally 90° or 120° for concrete or strap saddles. The stress is practically independent of the
width of the support provided it is small compared with the pipe
diameter. To this local stress at the saddle must be added the maximum longitudinal bending stress at the support for the pipe acting as
a continuous beam, unless a joint occurs at the support.

Ring Girders

Steel pipes may be laid above ground over ravines, marshy
ground etc. Buckling of the pipe and stresses in these circumstances
may be more severe than under the ground, as there is no sidefill
support and the pipe saddles cause stress concentrations. Rigid ring
girders at each support are useful in these circumstances. The rings
may be plain rectangular in cross section, or T or H shaped, with
the back fixed all round the pipe. The local pipe bending stress in
the vicinity of the ring may be evaluated using the ring theory
developed in Chapter 8. The longitudinal bending stress in the pipe
wall under a ring is

$$\frac{1.65}{0.91 + h\sqrt{hd/A}} \frac{Pd}{2h} \qquad (9.9)$$

where A is the cross sectional area of the ring. To this stress must be added any longitudinal stress in the pipe due to beam action between supports.

The ring stress in the ring girder is

$$\frac{1}{1 + 0.91A/(h\sqrt{hd})} \quad \frac{Pd}{2h} \tag{9.10}$$

In addition there will be local stresses due to the method of fixing the ring girder to the supports. Legs are securely fixed to the ring girder and rest on a sliding or roller bearing on a pier. The weight of the pipe and contents is transferred to the bottom half of the ring girder and thence through the legs to the pier.

A more comprehensive analysis of the stresses in ring girders is presented in Crocker and King (1967). See also Cates (1950) and Scharer (1933).

TEMPERATURE STRESSES

Pipes with high differences in temperature between inside and outside will be subject to radial and circumferential stresses due to temperature differences. The circumferential and longitudinal stresses have their maximum value on the inner and outer surfaces, being compressive on the inner surface and tensile on the outer surface if the temperature on the inner wall is above that on the outer wall, and the heat flow is steady. If the wall is thin in comparison with the pipe diameter, the stresses may be approximated by the following equations:

on the inner surface $F_c = F_\ell = -\dfrac{\alpha E T}{2(1-\nu)}$ (9.11)

and on the outer surface $F_c = F_\ell = \dfrac{\alpha E T}{2(1-\nu)}$ (9.12)

where α is the thermal coefficient of expansion, E is the modulus of elasticity, T is the temperature difference, ν is Poisson's ratio, F_c is circumferential stress and F_ℓ is longitudinal stress. The radial stress is compressive and is always less than the maximum circumferential and longitudinal stresses, being zero on both boundaries.

For normal climatic temperature differences the stresses are relatively low. Thus for T = 10°C, E = 200 000 N/mm², α = 12 $\times$ 10^{-6} per °C, ν = 0.3 for steel pipe, then F max = 17 N/mm².

A longitudinal stress may also be induced in a pipe restrained longitudinally as a result of temperature change after installation. The magnitude of this stress is αET (tensile if the temperature drops after installation) (see Ch. 10 for an analysis including effects of supports).

REFERENCES

Blair, J.S., 1946. Reinforecement of branch pipes. Engineering, 162.

Cates, W.H., 1950. Design standards for large diameter steel water pipe, J. Am. Water Works Assn., 42.

Crocker, S. and King, R.C., 1967. Piping Handbook, 5th Ed., McGraw Hill, N.Y.

Pearson, F.H., 1977. Beam behaviour of buried rigid pipelines. Proc. Am. Soc. Civil Engrs., 108 (EE5) 767.

Roark, R.J., 1954. Formulas for stress and strain, McGraw Hill, N.Y.

Rouse, H., 1961. Engineering Hydraulics, Wiley, N.Y., p. 418.

Scharer, H., 1933. Design of large pipelines. Trans. Am. Soc. Civil Engs., 98 (10).

Stephenson, D., 1971. Internal bracing for pipe confluences. Trans., S.A. Instn. Civil Engrs., 13 (11).

Sulzer, 1941. Patented stiffening collars on the branches of high-pressure pipelines for hydroelectric power works. Sulzer Technical Review, 2,10.

Süss, A. and Hassan, D.R., 1957. Reduction of the weight and loss of energy in distribution pipes for hydraulic power lants. Escher Wyss News, 30 (3) 25.

Swanson, H.S., Chapton, H.J., Wilkinson, W.J., King, C.L. and Nelson, E.D., 1955. Design of wye branches for steel pipes. J. Am. Water Works Assn., 47 (6) 581.

Timoshenko, S. and Goodier, J.N., 1951. Theory of Elasticity, McGraw Hill, N.Y.

LIST OF SYMBOLS

A	–	area of ring girder
a,b	–	horizontal and vertical axes of an ellipse
a,b,c	–	crotch plate top and side widths
D or d	–	pipe diameter
E	–	modulus of elasticity
f	–	permissible crotch plate stress
F	–	resultant force on crotch plate at any point, or stress
G	–	factor of safety
h	–	wall thickness

L	–	span or length
M	–	bending moment
P or p	–	water pressure (same units as f)
Q	–	saddle reaction
R or r	–	radius
S	–	stress applied to a plate
t	–	crotch plate thickness
T	–	temperature difference
u	–	distance from pipe wall to F
w	–	width of crotch plate at any point
W	–	load per unit length
X	–	horizontal distance from crown along centre-line of crotch plate
Y	–	vertical distance above mid-plane or crown of crotch plate
Z	–	section modulus
α	–	angle of confluence of branch pipe with main pipe
α	–	coefficient of temperature expansion
β	–	angle between crotch plate and main pipe axis, or angle of support
γ	–	specific weight
θ	–	angle of flare, or taper, of main pipe (measured from axis to cone)
φ	–	angle of flare, or taper, of branch pipe
ν	–	Poisson's ratio

CHAPTER 10

PIPES, FITTINGS AND APPURTENANCES

PIPE MATERIALS

Steel Pipe

Steel is one of the most versatile materials for pipe walls, as it is ductile yet has a high tensile strength. It is relatively easy to work, and the welded joint frequently used with steel is the strongest type of joint.

Steel grades used for pipes in the U.K., and their corresponding minimum yield stresses are as follows:

BS 4360	Grade 40	:	230 N/mm^2
	Grade 50	:	355 N/mm^2
	Grade 55	:	450 N/mm^2
US 572	Grade 42	:	290 N/mm^2
	Grade 50	:	345 N/mm^2
	Grade 60	:	414 N/mm^2
	Grade 65	:	448 N/mm^2

The higher grades are preferred for high-pressure pipelines but for low pressure pipelines the lower grade steels are more economical, they are easier to weld, and on account of the extra wall thickness they are more resistant to external loads.

Small-bore (less than 450 mm) steel pipes are rolled without seams, but large-bore pipes are made from steel plate, bent and welded either horizontally or spirally. Pipes are made in lengths up to 10 m and more and jointed on site. Steel pipe is more expensive than concrete, asbestos cement and plastic pipe for small bores and low pressures, and it requires coating and sheathing to prevent corrosion.

Cast Iron Pipe

Cast iron is more corrosion-resistant than steel, but more expensive and more rigid. In fact actual sand casting is now rarely used for plain pipe, although it is used for 'specials' such as bends, tapers and flanges. Plain pipes are normally formed of grey iron or ductile iron by centrifugal spinning. Standard pressure classes and dimensions of ductile iron pipes are specified in BS 4772 and of grey iron pipes in BS4622.

Asbestos Cement Pipe

Asbestos cement pipe is made of cement and asbestos fibre, which is able to resist relatively high tensile stresses. Although asbestos cement is relatively cheap, strong and corrosion-resistant, it is susceptible to shock damage and cannot be used for 'specials'.

Joints are made with a sleeve fitted with rubber sealing rings.

Concrete Pipe

Reinforced or prestressed concrete pipes are suitable and economical for large diameters. They are able to resist external buckling loads easier than steel on account of their extra wall thickness, and are corrosion resistant. Their main weaknesses appear to be in jointing and making later connections. There is as yet not much long-term experience with prestressed concrete pipes.

Plastic Pipe

Technological advances in plastics in the 1960's were rapid and this has been of special interest in pipeline engineering. Unplasticized polyvinylchloride pipes are now used extensively for chemicals, gases, drains, waterpipes, irrigation and sewer pipes in the United Kingdom and Europe. Polyethylene is also used to a limited extent for small bore pipes although it is usually more expensive than UPVC for large diameters. High density polyethylene has recently been used with some success for large diameter pipelines.

Glass fibre and resin is also used to a limited extent for pipelines. It is comparatively expensive but used with an inert filler it is suitable for some specialised applications as it is corrosion-resistant, rigid, light and strong.

Although plastic pipe has not yet stood the rigorous test of time and is still not acceptable under many by-laws, its many advantages are causing it to gain rapid acceptance amongst engineers.

The working stress of plastic is less 14 N/mm^2 so plastic cannot be used for large diameter high-pressure mains. It is also subject to limited strength deterioration with time and its flexibility can cause buckling and collapse. It is resistant to many corrosive fluids and can be used in difficult locations such as undersea outfalls and gas pipes. The friction loss is lower than for most materials and it can be used beneficially as a lining to other types of pipes. It is less subject to encrustation than other types of pipe, accommodates greater ground movements and is easier and lighter to lay. On the other hand it has a high coefficient of thermal expansion, so joints may be loosened if the pipe contracts on cooling. It is susceptible to damage and may distort or even collapse under load. Ribbed and stiffened pipes are being developed to overcome these limitations.

Plastic piping is normally made of thermoplastic which is relatively easy to extrude when heated. Polyethylene and PVC are thermoplastics. As PVC is flexible and exhibits creep, it is generally used in an unplasticized form (UPVC) for the manufacture of pipes. Polyethylene is easier to work than UPVC and for this reason earlier developments were with polyethylene. With improvement in production techniques UPVC is now cheaper than polyethylene and it holds certain other advantages. It is stronger and has a smaller coefficient of thermal expansion than polyethylene, although recently-developed high-density polyethylene (HDPE) has improved properties. Polyethylene is more ductile than UPVC and for this reason is often preferred for gas piping. Rubber toughened plastics such as ABS have also been developed for flexible pipework. Properties of plastics are tabulated in the appendix. Because of the low elasticity of plastics, water hammer pressures are less severe than for other materials.

Plastic pipes may be jointed by solvents of fusion welding, although these methods are difficult for large diameters. Screwed or factory fitted flanges or spigots and sockets with rubber rings are also used.

The manufacture of moulded bends, tees and branches is difficult in UPVC and even more so in HDPE so steel or C.I. fittings are frequently used.

The price of PVC pipes in the U.K. dropped by more than 30% in the 1960's, whereas other pipe materials are continually increasing in cost. In the U.K., PVC pipes are cheaper than asbestos cement and cast iron pipes for diameters less than 450 mm and pressures less than 1 N/mm². It is rapidly replacing salt-glazed and asbestos-cement pipes for sewers and low pressure systems.

At least £700 million was spent in 1972 on plastic piping throughout the world. The figure is likely to rise considerably as plastic pipes are used in ever increasing proportions and new discoveries may yet revolutionize pipeline engineering (see also Boucher, 1948; BVMA, 1964; Paul, 1954).

LINE VALVES

There are many types of valves for use in pipelines, the choice of which depends on the duty. The spacing of valves and the size will depend on economics. Normally valves are sized slightly smaller than the pipe diameter, and installed with a reducer on either side. In waterworks practice it is preferable to keep the soffit of the valve at the same level as the soffit of the pipe, to prevent air being trapped, whereas in sewerage and solids transport, the inverts should be lined up. In choosing the size, the cost of the valve should be weighed against the cost of the head loss through it, although in certain circumstances it may be desirable to maintain the full pipe bore (to prevent erosion or blockage).

Isolating valves are frequently installed at intervals of 1 to 5 km, the spacing being a function of economics and operating problems. Sections of the pipeline may have to be isolated to repair leaks and the volume of water which would have to be drained to waste would be a function of the spacing of isolating valves. Scour valves are installed at the bottom of each major dip in the pipeline profile.

It is sometimes advisable to install small-diameter by-pass valves around in-line valves to equalize pressures across the gate and thus facilitate opening (which may be manual or by means of an electric or mechanical actuator).

Sluice Valves

Sluice valves, or gate valves, are the normal type of valves used for isolating or scouring. They seal well under high pressures and when fully open offer little resistance to fluid flow.

There are two types of spindles for raising the gate: A rising spindle which is attached to the gate and does not rotate with the handwheel, and a non-raising spindle which is rotated in a screwed attachment in the gate (Fig. 10.1). The rising spindle type is easy to lubricate.

The gate may be parallel-sided or wedge-shaped. the wedge-gate seals best but may be damaged by grit. For low pressures resilient or gunmetal sealing faces may be used but for high pressures stainless steel seals are preferred.

Despite sluice valves' simplicity and positive action, they are sometimes troublesome to operate. They need a big force to unseat them against a high unbalanced pressure, and large valves take many minutes to turn open or closed. Some of the problems can be overcome by installing a valve with a smaller bore than the pipeline diameter.

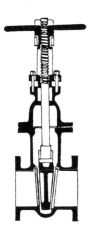

Fig. 10.1   Sluice valve

Butterfly Valves (Fig. 10.2)

Butterfly valves are cheaper than sluice valves for larger sizes and occupy less space. The sealing is sometimes not as effective as for sluice valves, especially at high pressures. They also offer a fairly high resistance to flow even in the fully open state,   because the thickness of the disc obstructs the flow   even when it is rotated

90 degrees to the fully open position. Butterfly valves, as well as sluice valves, are not suited for operation in partly open positions as the gates and seatings would erode rapidly. As both types require high torques to open them against high pressure, they often have geared handwheels or power driven actuators.

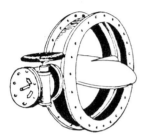

Fig. 10.2   Butterfly valve

Globe Valves

Globe valves have a circular seal connected axially to a vertical spindle and handwheel. The seating is a ring perpendicular to the pipe axis. The flow changes direction through 90 degrees twice thus resulting in high head losses. The valves are normally used in small-bore pipework and as taps, although a variation is used as a control valve.

Needle and Control Valves (Fig. 10.3)

Needle valves are more expensive than sluice and butterfly valves but are well suited for throttling flow. They have a gradual throttl-ing action as they close whereas sluice and butterfly valves offer little flow resistance until practically shut and may suffer cavitation damage. Needle valves may be used with counterbalance weights, springs, accumulators or actuators to maintain constant pressure conditions either upstream or downstream of the valve, or to maintain a constant flow. They are streamlined in design and resistant to wear even at high flow velocities. The method of sealing is to push an axial needle or spear-shaped cone into a seat. There is often a pilot needle which operates first to balance the heads before opening.

A valve with a similar hydraulic characteristic to the needle valve is the sleeve valve which is suitable for use discharging into the atmosphere or an open bay. They seal by a sleeve which slides over orifices around the tube and spread the flow in an umbrella-shaped spray, thereby dissipating the energy. Needle and sleeve valves are also occasionally used as water hammer release valves when coupled to an electric or hydraulic actuator (see Chapter 4). They require dismantling for maintenance as the working parts are inside the valve. The cone valve is a variation of the needle valve but the sealing cone rotates away from the pipe axis instead of being withdrawn axially.

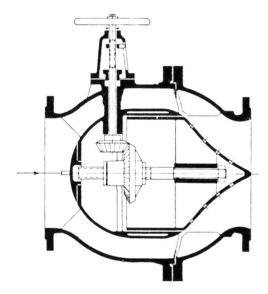

Fig. 10.3   Needle valve

Spherical Valves (Fig. 10.4)

Spherical valves have a rotary plug with an axial hole through it. When the valve is fully opened there is no resistance to flow as the bore is equal to that of the pipe. The valves close by rotating the sphere, and normally have an offset action to unseat them before rotating the sphere into the open position.

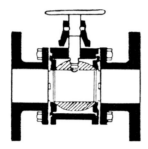

Fig. 10.4   Spherical valve

Reflux Valves   (Fig. 10.5)

Reflux, or non-return, or check valves as they are also known, are used to stop flow automatically in the reverse direction. Under normal flow conditions the gate is kept open by the flow, and when the flow stops, the horizontally hinged gate closes by gravity or with the aid of springs. A counterbalance can be fitted to the gate spindle to keep the gate fully open at practically all flows, or it could be used to assist in rapid closing when the flow does stop. Springs are also sometimes used to assist closing. Larger reflux valves may have multi-gates, in which case the thickness of each gate will partially obstruct the flow, and swabbing of the pipeline may be difficult. Mounted in horizontal pipes, the gates of some types of reflux valves tend to flutter at low flows, and for this reason an offset hinge which clicks the gate open is sometimes used.

AIR   VALVES   (see also Lescovich, 1972; Parmakian, 1950; Sweeton,
        1943).

Two types of air release valves are normally used in pipelines. One is a small-orifice automatic air release valve and the other is a large-orifice air vent valve.

Air Vent Valves

When a pipeline is filled, air could be trapped at peaks along the profile, thereby increasing head losses and reducing the capacity of the pipeline. Air vent valves are normally installed at peaks to permit air in the pipe to escape when displaced by the fluid. They also let air into the pipeline during scouring or when emptying the pipeline. Without them vacuum may occur at peaks and the pipe could collapse, or it may not be possible to drain the pipeline completely. It is also undesirable to have air pockets in the pipe as they may cause water hammer pressure fluctuations during operation of the pipeline. The sealing element is a buoyant ball which, when the pipe is full, is seated against an opening at the top of the valve. When the pressure inside the pipe falls below the external pressure the ball drops thereby permitting air to be drawn into the pipe. When the pipe is being filled the valve will remain open until the water fills the pipe and lifts the ball against the seating. The valve will not open again until the pressure falls below the external pressure.

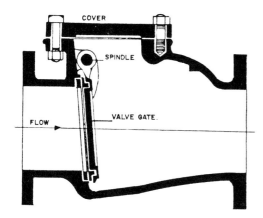

Fig. 10.5   Reflux valve

The valve will tend to blow closed at high air velocities. It should be realised that a differential pressure of only 0.1 N/mm² (13 psi) will cause sonic velocity of air through the valve (300m/s). Pressure differences around 0.027 N/mm² (4 psi) have been found to close some air vent valves. The slamming closed may damage the hollow ball seal.

In practice a diameter approximately 1/10th of the pipe diameter is used, although larger diameters and duplicate installations are preferable. The valve is referred to by the size of the inlet connection diameter, and the orifice diameter is usally slightly smaller.

The size of air vent valve will depend on the rate of filling or desired scouring rate of the pipe. The volume rate of flow of air through an orifice is approximately 40 times the flow of water for the same pressure difference, but full vacuum pressure should not be allowed to develop.

Air vent valves should be installed at peaks in the pipeline, both relative to the horizontal and relative to the hydraulic gradient. Various possible hydraulic gradients including reverse gradients during scouring, should be considered. They are normally fitted in combination with an air release valve, as discussed in the next section.

Equations for sizing air valves are derived in Chapter 5. As a rule of thumb the orifice area in m^2 should be Q/100 where Q is the rate of flow of air in m^3/s at initial pressure. The rate of flow of free air is the most difficult factor to determine though. During fitting operations, the rate of evacuation of air through all vent valves will equal the rate of filling the line. Care should be taken as all the air is evacuated as the flow of water will suddenly be stopped and water hammer is possible.

Air Release Valves

Air is entrained in water in many ways; by vortices in the pump suction reservoir or merely by absorption at exposed surfaces, at air pockets or in the suction reservoir. The air may be released when there is a drop in pressure, either along a rising main or where the velocity is increased through a restriction such as a partially closed valve. An increase in temperature will also cause air to be released from solution. Small orifice air release valves are installed on the pipeline to bleed off the air which comes out of solution.

Small orifice air valves are designated by their inlet connection size, usually 12 to 50 mm diameter. This has nothing to do with the air release orifice size which may be from 1 to 10 mm diameter. The larger the pressure in the pipeline, the smaller need be the orifice size. The volume of air to be released will be a function of the air entrained which is on the average 2% of the volume of water (at atmospheric pressure), (see Chapter 5).

The small orifice release valves are sealed by a floating ball, or needle which is attached to a float. When a certain amount of air has accumulated in the connection on top of the pipe, the ball will drop or the needle valve will open and release the air. Small orifice release valves are often combined with large orifice air vent valves on a common connection on top of the pipe. The arrangement is called a double air valve, (see Fig. 10.6). An isolating sluice valve is normally fitted between the pipe and the air valves.

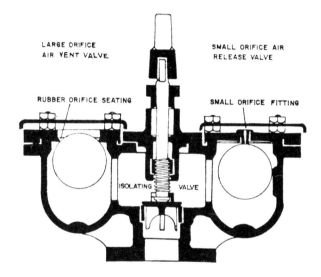

Fig. 10.6   Double air valve

Double air valves should be installed at peaks in the pipeline, both with respect to the horizontal and the maximum hydraulic gradient. They should also be installed at the ends and intermediate points along a length of pipeline which is parallel to the hydraulic grade line. It should be borne in mind that air may be dragged along in the direction of flow in the pipeline so may even accumulate in sections falling slowly in relation to the hydraulic gradient. Double air valves should be fitted every $\frac{1}{2}$ to 1 km along descending sections, especially at points where the pipe dips steeply.

Air release valves should also be installed on all long ascending lengths of pipeline where air is likely to be released from solution due to the lowering of the pressure, again especially at points of decrease in gradient. Other places air valves are required are on the discharge side of pumps and at high points on large valves and upstream of orifice plates and reducing tapers.

THRUST BLOCKS (See also Morrison, 1969)

Unbalanced thrust results at a bend in a pipeline due to two actions:

(1)   The dynamic thrust due to the change in direction of flow. The force in any direction is proportional to the change in momentum in that direction and is:

$$F_{x1} = \rho q \Delta V_x \qquad (10.1)$$

where F is the force, ρ is the fluid mass density, q is the flow and ΔV_x the reduction of the component of velocity in the direction x. This force is, under normal conditions, negligible compared with the force due to the internal pressure in the pipe.

(2)   The thrust in the direction of each leg of the bend due to the pressure in the pipe is:

$$F_{x2} = pA \qquad (10.2)$$

where p is the internal pressure, A is the cross sectional area of pipe flow and θ is the angle of deviation of the pipe (Fig. 10.7). The resultant outward thrust is the vector sum of the forces in both directions of the pipe axis, and is

$$F_r = 2pA \sin \theta /2 \qquad (10.3)$$

The unbalanced thrust may be counteracted by longitudinal tension in an all-welded pipeline, or by a concrete thrust block bearing against the foundation material. In the case of a jointed pipeline the size of the block may be calculated using soil mechanics theory. In addition to frictional resistance on the bottom of the thrust block and the circumference of the pipeline, there is a lateral resistance against the outer face of the pipe and block. The maximum resisting pressure a soil mass will offer is termed the passive resistance and is (Capper and Cassie, 1969)

$$f_p = \gamma_s h \frac{1 + \sin\phi}{1 - \sin\phi} + 2c\sqrt{\frac{1 + \sin\phi}{1 - \sin\phi}} \qquad (10.4)$$

where f_p is the resisting pressure at depth h, γ_s is the soil density, ϕ is the effective internal angle of friction of the soil and c is its cohesion. If the thrust block extends from the surface to a depth H below the surface, and its length is L, the total resisting thrust is

$$F_p = \gamma_s \frac{H^2}{2} L \frac{1 + \sin\phi}{1 - \sin\phi} + 2cHL\sqrt{\frac{1 + \sin\phi}{1 - \sin\phi}} \qquad (10.5)$$

This maximum possible resistance will only be developed if the thrust block is able to move into the soil mass slightly. The corresponding maximum soil pressure is termed the passive pressure. The minimum pressure which may occur on the thrust block is the active pressure, which may be developed if the thrust block were free to yield away from the soil mass. This is

$$f_a = \gamma_s h \frac{1 - \sin\phi}{1 + \sin\phi} - 2c\sqrt{\frac{1 - \sin\phi}{1 + \sin\phi}} \qquad (10.6)$$

which is considerably less than the passive pressure and will only be developed if the force on which it is acting is free to move away from the soil exerting the pressure.

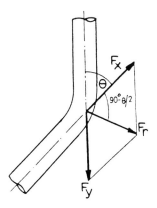

Fig. 10.7   Thrust at a bend

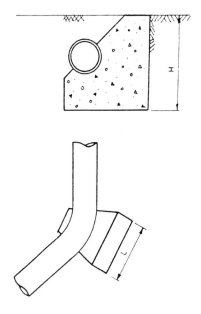

Fig. 10.8   Thrust block

In practice the pressure which may be relied upon is slightly less
than the passive pressure, as full movement of the pipe is usually
not permissible. A factor of safety of at least 2 should be used with
the passive resistance formula. The inward movement of the top of a
thrust block to activate full passive pressure varies from H/20 for
soft clays to H/200 for dense sand, (CP 1972), where H is the depth
of the excavation. These movements are seldom permissible, and the
"at rest" coefficients are probably more realistic. The "at rest" ratio
of horizontal soil stress to vertical soil stress varies from 0.4 for
loose sands to 0.7 for very plastic clays. An impact factor should be
used in allowing for water hammer pressures. Typical values of angle
of internal friction ϕ and cohesion c for various soils are indicated
in Table 10.1.

TABLE 10.1   Strength of Soils – typical values

Type of soil	Angle of friction ϕ	Cohesion N/mm^2
Gravel	35°	0
Sand	30°	0
Silt	28°	0.007
Dense clay	5°	0.035
Soft saturated clay	0	0.15

Example

Calculate the size of a thrust block to resist the unbalanced pressures at a 45° bend in a 1 000 mm diameter pipe operating at a pressure of 2.0 N/mm^2. Length of 1 pipe length = 10 m, depth = 1 m, ϕ = 30°, C = 0.005 N/mm^2. Lateral force F_r = 2 × 2.0 $\frac{\pi}{4}$ × -1^2. sin 22$\frac{1}{2}$° = 1.20 MN.

Try a thrust block 3×3×3 m:

Weight of thrust block 3×3×3×2 200×9.8/10^3	=	600kN
Weight of 10 m length of pipe	=	40kN
Weight of water in pipe 0.785×1^2×10×9.8×1 000/10^3	=	80kN
Weight of soil above pipe 1×1×7×1 800×9.8/10^3	=	130kN
Total weight	=	850kN
Frictional resistance 0.3×850	=	250kN

Lateral resistance of soil against block:

$$\frac{1 + \sin\phi}{1 - \sin\phi} = \frac{1 + 0.5}{1 - 0.5} = 3$$

F = 18 000×$\frac{3^2}{2}$×3×3/10^3 +2×5×3×3 $\sqrt{}$ 3   =   889kN

Lateral resistance of soil against projection of pipe:

(18 000×1.5×3/10^3 ×2×5 $\sqrt{3}$) 1×6.5   =   640kN

Total lateral resistance   = 1770kN

Factor of safety 1.77/1.20 = 1.48

A thrust block should be designed so that the line of action of the resultant of the resisting forces coincides with the line of thrust of the pipe. This will prevent overturning or unbalanced stresses. This may best be done graphically or by taking moments about the centre of the pipe.

194

Thrust blocks are needed not only at changes in vertical or horizontal alignment of the pipeline, but also at fittings which may not be able to transmit longitudinal forces, such as flexible couplings. Viking Johnson or similar couplings are frequently installed on one side of a valve in a valve chamber, to facilitate installation and removal of the valve. The opposite wall of the valve chamber would thus have to be designed as a thrust block.

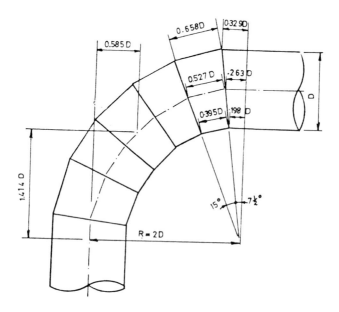

Fig. 10.9   Segment geometry : standard medium-radius 90° fabricated bend.

FORCES INDUCED BY SUPPORTS

The interaction of a pipe and its supports is a complex subject. There are many secondary and tertiary forces and displacements involved. The effects are generally greatest in pipes supported above ground, on plinths or by other isolated supports. There is a tendency of the pipe to expand or contract longitudinally and circumferentially under the effects of temperature change, differential temperatures, secondary strains due to the Poisson ratio effect, and partial or total fixity of supports which impose forces on the pipe vertically, longitudinally and laterally. No support is completely rigid and the more flexible the support the less force or reaction it imposes on the pipe.

The stresses on the pipe may be loosely classified as follows:

<u>Primary</u>; circumferential due to internal pressure across diameter, longitudinal due to internal pressure across the cross section.

<u>Secondary</u>; bending at elbows, branches, changes in section and across supports. Temperature effects. Poisson ratio counter-strains at right angle to primary strain. Due to reaction of supports.

<u>Tertiary</u>; lateral deflection due to longitudinal expansion.

Although the secondary and tertiary stresses are often not accounted for in pipe design, they can be very significant.

Longitudinal Stress

If the end of a pipe is blanked off and unheld the longitudinal stress in the pipe wall is

$$f = \frac{\pi d^2 p/4}{\pi d t} = \frac{1}{2}\frac{pd}{2t} \tag{10.7}$$

i.e. $\frac{1}{2}$ the circumferential stress.

The Poisson ratio effects tries to reduce the length by $\nu Lpd/2tE$ and if the ends are held the wall tensile stress becomes $Ee = \nu pd/2t$ and corresponding force is $\nu p \pi d^2/2$ where E is the elastic modulus, t the wall thickness, e the strain and ν is Poisson's ratio (about 0.3 for steel).

Temperature stresses

A pipe will expand or contract by $\alpha \Delta T$ where α is the coefficient of expansion and ΔT the temperature change in time.

A pipe restrained longitudinally will undergo a stress $\alpha E \Delta T$ where the coefficient of expansion for steel, α, is 12×10^{-6} per °C and for plastic, 50×10^{-6}.

If there is a temperature gradient across the pipe wall a circumferential stress is created equal to $\pm \alpha E \Delta T/2(1-\nu)$.

Forces at Bends

The force due to water pressure along each axis at a pipe bend is pA. The net resultant force acting outwards is 2pA sin $(\theta/2)$ where θ is the bend angle. If the resultant force is not countered with a thrust block it creates a longitudinal force in the pipes equal to 2pA $\sin^2 (\theta/2)$.

Lateral Movement

A straight pipe of length x fixed at both ends will buckle outwards a relatively big amount under expansion due to temperature etc., by an amount

$$dy \cong \sqrt{x \cdot dx/2} \tag{10.8}$$

since $dy^2 + (x/2)^2 = (x/2 + dx/2)^2$

$$dy = \sqrt{2(x/2)\ (dx/2)} \quad = \quad \sqrt{x \cdot dx/2}$$

e.g.   x = 100m, dx = $x\alpha\Delta T$ = 100 × 12 × 10^{-6} × 20 = 0.02m

$$dy = \sqrt{100 \times 0.02/2} = 1m.$$

If the pipe is restrained laterally it could try to buckle between supports or twist supports or exert a lateral force equal to that required to prevent the buckling. For example, the deflection of a simple supported (laterally) pipe beam, dy, under a force F is

$$dy = \frac{1}{48} FL^3/EI$$

= 1   for above example.

$$\therefore F = 1 \times 210 \times 10^9 \times \frac{\pi}{8} \times 1^3 \times .01/(100^3 \times \frac{1}{48})$$

= 40 kN

The lateral deflection is referred to as snaking and can cause pipes to fall off the side of supports, or bend the supports.

A pipe fixed at both ends with an elbow will bend outwards under expansion by an amount:

dy = dx/(y/a + y/b)

where dx is the net expansion in length of pipe with length (a + b) = c due to temperature, pressure etc.

Forces on Supports

The forces on the support blocks or pillars or hangers will be equal and opposite to the forces exerted by the supports on the pipe.

Free to Slide: The longitudinal force of a freely sliding or highly held pipe equals the friction force on the block. This is μW, where μ is the friction coefficient and W is the normal force (weight) of the pipe on the block. There is also a lateral force to resist snaking.

Fixed to Block: The resulting force is the smaller of (i) maximum force which can be imposed on block by pipe e.g. friction between pipe and block, or (ii) sliding resistance of block, or (iii) resistance of support at the other end, or (iv) maximum net total longitudinal force in pipe due to temperature change, Poisson ratio effect and bends.

There is thus a distinction between the free to slide case and the fixed case in that (i) or (ii) will apply to a "free to slide" pipe and case (iii) and (iv) will generally apply to a pipe rigidly fixed to the support.

Thus if one end of a pipe is free (see Fig. 10.9) then there will obviously be no force at that end, and beyond the first support the force is computed as above, and beyond the second support also as above etc. For the first few lengths case (i) or (ii) may apply and then case (iii) or (iv) will take over as one of these is less. For case (iv) the "force at the other end" may in fact be that due to other support blocks.

Unbalanced Forces

Generally it can be expected that forces in welded pipework balance out. That is forces in one direction at one end are equal and opposite to those at the other end. Along the length of the pipe however there may be bending moments. For instance at elbows or bends there are net forces of liquid on the pipe wall inside acting to try to straighten the bend.

During transient flow conditions there may however be large unbalanced forces which could pull the pipe off supports. For instance following closure of a vapour pocket in a pipe water hammer pressure at one end of a line may force the pipe in one direction. Although the pressure wave may travel along the pipe length and eventually re-balance forces, there are momentary unbalanced forces which can cause significant movement starting at bends.

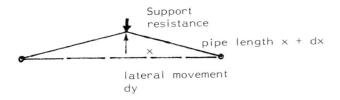

Fig. 10.10   Lateral movement due to temperature expansion

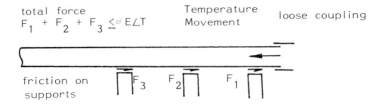

Fig. 10.11   Longitudinal forces in pipe wall created by friction
             on supports

FLOW MEASUREMENT

Venturi Meters   (Fig. 10.12)

The Venturi meter offers an accurate method of flow measurement with minimum head loss. It is used mainly for large-diameter pipelines. The venturi throat has a converging section and gradually diverging section to minimize head losses.

Flow measurement is based on the Bernoulli equation which is rearranged as

$$\frac{V_2^2}{2g} - \frac{V_1^2}{2g} = h_1 - h_2 = h \qquad (10.9)$$

Solving for discharge

$$Q = C_d A_2 \sqrt{2g\,\Delta h} \qquad (10.10)$$

where C_d is a discharge coefficient which equals

$$C_v / [1 - (\frac{d_2}{d_1})^4]^{\frac{1}{2}} \qquad (10.11)$$

where d is the diameter and C_v is a velocity coefficient which allows for flow separation at the throat and includes a factor for upstream and downstream conditions in the pipe. A venturi meter should be calibrated in place and should have a straight length of at least 5 pipe diameters for very small throat/pipe diameter ratios, increasing to 10 or even 20 pipe diameters for larger throats, or from branches or double bends, is required (BS 1042). Full bore branches, partly closed valves and fittings which impart swirling motion may need up to 100 pipe lengths downstream before a meter. The head loss in a venturi meter is only approximately $0.1V_2^2/2g$.

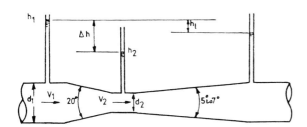

Fig. 10.12 Venturi meter

Nozzles

A contracted form of venturi meter is the nozzle meter, which has a short bellmouth rounded inlet and abrupt expansion beyond it. The discharge equation is similar to that for the venturi meter but the velocity head recovery beyond is very small.

Orifices

An orifice consisting of a thin plate with a central orifice is a popular method of measuring flow in large pipes. There is an appreciable orifice contraction coefficient which may be as low as 0.61. The discharge formula is similar to Equ. 10.10 where

$$C_d = C_v C_c / \sqrt{1 - C_c^2 (\frac{d_2}{d_1})^4}$$

(10.12)

The velocity head recovery is low but the orifice has the advantage that it is short and can be installed in a short length of straight pipe. The lengths of straight pipe required on each side of the orifice are similar to those for venturi meters. The orifice should be calibrated in place for accurate results.

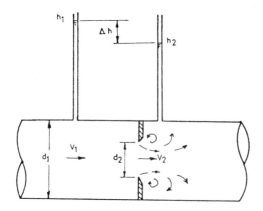

Fig. 10.13 Orifice meter

Bend Meters

A greater pressure is exerted on the outside of a bend in a pipe than on the inside, due to centrifugal force. The pressure difference can be used to measure the flow. The meter should be calibrated in place, but the discharge coefficient for Equ. 10.10 is approximately

$$C_d = r/2d$$

(10.13)

where r is the radius of curvature of the centreline.

Mechanical Meters

Flow to water consumers is invariably measured by mechanical meters which integrate the flow over a period of time. A common type has a rotating tilted disc with a spindle which rotates a dial. Other types have rotating wheels, lobes or propellers.

There is also a type of meter which reacts to the drag on a deflected vane immersed in the flow. The meter records for partly full conduits as well.

The rotometer consists of a calibrated vertical glass tube which has a taper increasing in diameter upwards. A float is suspended in the upward flow through the tube. The float positions itself so that the drag on it (which depends on the flow past it and the tube diameter) equals the submerged weight.

The object may be either a sphere or a tapered shape with vanes to make it rotate and centre itself in the tube.

Electromagnetic Induction

By creating a magnetic field around a pipe of non-conducting material and ionizing the liquid by inserting electrodes, an electro-motive force is induced and can be measured. The method has the advantage that there is no loss of head and a variety of liquids including sewage can be measured by this means. A similar tech-nique which also does not obstruct the flow and is based on sonic velocity of an impressed shock wave, is also being developed. The difference between upstream and downstream sonic velocity is measured.

Mass and Volume Measurement

The most accurate methods of measurement of flow are by mass or volumetrically. To measure the mass or weight of fluid flowing over a certain time the flow is diverted into a weighing tank. The volume flow may be measured from the volume filled in a certain time.

TELEMETRY

Automatic data transmission and control is often used on pipeline systems. Telemetry systems are tending to replace manual data hand-ling which is more expensive and less reliable than automatic methods. Cables may be laid simultaneously with the pipeline.

Data which may be required include water levels in reservoirs, pressures, flows, opening of valves, temperature, quality or pro-

perties of the fluid, pump speed or power consumption. The information is read by a standard instrument linked by a mechanical
device to a local dial or chart, or to a coder and transmitter.

Signals may be sent out from a transmitter in the form of continuous analogue waves or in a series of digits. Analogue messages
are usually less accurate (up to 2 percent accuracy) but also less
expensive than digital. The transmitter may repeat messages continuously or be interrogated at regular intervals, termed scanning.
With digital systems data are converted to binary code electric
pulses by relays, magnetic cores or contacts. Every possible level
of the measurand is represented by a definite combination of binary
coded numbers. Signals are transmitted in series and in groups
known as bits. The signals may be transmitted via telephone lines
or private multiplex cables. One pair of conductors is needed for
each bit, plus a common feed. A number of units of data may be
sent in rotation via the same conductors.

Alternatively hydraulic or pneumatic signals may be conveyed
over short distances by pipe. Radio transmission becomes economic
for systems with long transmission distances.

The transmission system will send the messages to a receiver
linked to a decoder or digital comparater. Signals can be decoded
from analogue to digital and vice versa. The decoder feeds the
information to a data bank or by means of a servo motor can dis-
play the readings on dials, charts or lights or sound alarms at
extreme values of the measurand. Data is frequently displayed on
a mimic diagram (a diagramatic picture) of the system with lights
or dials at the position of the unit being observed. These diagrams
are mounted in central control rooms.

The information may also be fed to a data storage system and
stored on paper or magnetic tapes, discs or cards. The information
may be fed directly or at a later stage to a computer to be used
for decision-making or manipulating the data. Large computers can
be programmed to optimize a system, for instance by selecting those
pumps which would result in minimum pumping cost, or turbines
which would match power demand. The computer could send signals
to close valves or operate hydraulic machinery or water hammer
relief mechanisms.

Mini digital computers are now the most popular form of control of telemetered systems. They are slightly cheaper than analogue computers but considerably cheaper than larger digital computers. They can be programmed to perform certain tasks or computations automatically but cannot be used easily for other computations for which they are not programmed. They can be increased in capacity or linked to larger computers as the need arises.

REFERENCES

Boucher, P.L., 1948. Choosing valves, Kilmarnock.
British Valve Manufacturers Assn. 1964. Valves for the Control of Fluids.
BS 1042, Flow measurement, BSI, London.
Capper, P.L. and Cassie, W.F., 1969. The Mechanics of Engineering Soils, 5th Ed., Spon, London.
CP 1972. Lateral Support in Surface Excavations, S.A. Instn. Civil Engs.
Lescovich, J.E., 1972. Locating and sizing air release valves. J. Am. Water Works Assn., 64 (7).
Morrison, E.B., 1969. Nomograph for the design of thrust blocks, Civil Engg., Am. Soc. Civil Eng.
Paul, L., 1954. Selection of valves for water services. J. Am. Water Works Assn., 46 (11) 1057.
Parmakian, J., 1950. Air inlet valves for steel pipelines. Trans. Am. Soc. Civil Engs., 115 (438).
Sweeten, A.E., 1943. Air inlet valve design for pipelines. Engg. News Record, 122 (37).

LIST OF SYMBOLS

A	–	cross sectional area
c	–	soil cohesion
C_c	–	contraction coefficient
C_d	–	discharge coefficient
C_v	–	velocity coefficient
d	–	internal diameter
e	–	strain
E	–	elastic modulus
f	–	stress
F	–	force
g	–	gravitational acceleration
h or H	–	depth below surface

p	–	fluid pressure
q	–	flow rate
r	–	radius of curvature
t	–	wall thickness
T	–	temperature
V	–	velocity
W	–	normal force
x	–	length
x,y	–	directions
y	–	lateral displacement
α	–	temperature coefficient
γ_s	–	soil density
μ	–	friction coefficient
ν	–	Poisson's ratio
ρ	–	soil mass density
ϕ	–	angle of internal friction of soil

CHAPTER 11

LAYING AND PROTECTION

SELECTING A ROUTE

The selection of a suitable route for a pipeline has an important bearing on the capital cost and operating costs. A pipeline route is selected from aerial photos, topographical and cadastral plans, on-site inspections and any other data available on the terrain, obstacles and local services. In selecting a route, the costs and prcticability have to be considered. Care should be taken to ensure the ground profile is below the hydraulic grade line. (Low-flow conditions should be considered as well as peak rates, as the hydraulic gradient is flattest for low flows). If there were a peak above the grade line between the input and discharge heads, obviously pumps would have to be designed to pump over this peak. Peaks may also be points of possible water column separation which result in water hammer overpressures. On the other hand the general level of the pipeline route should be kept as near to the hydraulic grade lines as possible to minimize pressures and consequently pipe costs.

Once a preliminary route is selected, it is pegged out by a surveyor. Pegs are put in along the centre-line at 10 to 100 m regular intervals and at changes in grade and at horizontal deflec-tions. Offset pegs, 2 to 5 m from the centre line pegs are also put in for use during laying. The levels of pegs are observed and the profile is then plotted in the drawing office, with ground levels, bed levels and depth indicated at each peg. Holes may be augered along the route to identify materials which will have to be excavated. Strength tests may be done on soil samples to decide whether trench shoring will be necessary.

Underground and overhead services should be accurately located at this stage or at least before excavation to avoid them. (Drains and underground cables may have to be supported on bridges, or a heading could be hand-excavated under them).

LAYING AND TRENCHING

The type of pipe to be used for any job will largely be a matter
of economics, although the facilities for laying, life of the pipe
materials and other factors affect the decision. The lengths of
delivered pipes will depend on the type of pipe, the weight, the
rigidity, size of transport vehicles and method of jointing. Concrete
pipes are often supplied in 2 to 4 m lengths, while steel pipes may
be over 10 or even 20 m long, and thin walled plastic pipes may
be supplied in rolls. Pipes should be handled carefully to avoid
damage to coatings or pipe or distortion. They should be stored
and transported on padded cradles or sandbags. Pipes may be lowered into the trench by belt slings from mechanical booms or tripods.
For small diameter thick walled welded pipes, jointing and wrapping
may be done at the side of the trench, and the pipe then 'snaked'
into the trench. The working width required for construction of pipelines varies with the method of excavation and laying. If the
excavation and laying is done by hand, 3 m may be wide enough
for small diameter pipes. On the other hand for large mains where
excavation, laying and possibly even site coating is done mechanic-
ally, working widths up to 50 m may be required. Reserve widths
of 20 to 30 m are common, and allow for one or two additional pipelines at a later stage.

Once the trench is excavated, sight rails may be set up across
the trench with the centre line of the pipe marked thereon. The
frames are set a definite height above the bed and the pipe levelled
with the assistance of these (BSCP, 1970).

Pipes are normally laid with a minimum cover of 0.9 m to 1.2
m to avoid damage by superimposed loads. A minimum cover of 1
m is necessary in the U.K. to prevent frost penetration. The depth
may be reduced in rocky or steep ground to minimize costs in which
case a concrete cover may be necessary. The width of trench is
normally 0.5 m to 0.8 m more than the pipe diameter, with local
widening and deepening for joints. (0.8 m is used for pipes over
300 mm diameter which are difficult to straddle while laying and
jointing).

The sides of the open trench will have to be supported in unco-hesive or wet soils and for deep trenches, for safety reasons. The shoring should be designed to resist the active lateral soil pressure. Alternatively the sides of the trench may be battered back to a safe angle. Spoil should be kept well away from the excavation to avoid slips. At least one trench depth away is advisable. Drains may be required in the bottom of a trench in wet ground to keep the trench dry while working, to reduce the possibility of embankment slips and to enable the backfill to be compacted properly. Drain pipes should be open jointed and bedded in a stone filter with frequent cross-drains to lead away water. In very wet ground cutoff drains beside the trench may be necessary.

Pipes may be bedded on the flat trench bottom, but a proper bedding is preferable to minimize deflections of flexible pipes or to reduce damage to rigid pipes. The bed of the trench may be pre-shaped to fit the profile of the pipe. In rocky ground a bed of sandy material may be built up and trimmed to shape. Beds of sand, gravel or crushed stone up to 20 mm in size, brought up to the haunches of the pipe, are often used, as they do not exhibit much settlement and are easy to compact. Concrete beds and haunches or surrounds are sometimes used for low-pressure rigid pipes and sewers. Bedding should be continuous under joints but haunches should stop short of joints to facilitate replacing damaged pipes. Fig. 11.1 shows typical types of bedding. (Refer to Chapter 7 to bedding factors associated with different types of bedding) (CPA, 1967; ACPA, 1970).

The best method of backfilling and compaction depends on the type of soil. Soil properties may be described in terms of the plastic limit (PI), liquid limit (LL) and plasticity index (PI = LL-PL). For soils with low PI's fill may often be satisfactorily compacted merely by inundation. For high LL soils, dynamic compaction is required i.e. stampers or vibrators. Backfill densities may be specified in terms of the Proctor Standard Density. For large pipes it is practice to compact the bed with a width of at least half the pipe diameter under the pipe to 95 to 100 % Proctor density. The soil fill around the pipe should be compacted in 100 to 150 mm layers to 90 or 95% Proctor density up to haunch level or to 3/4 of the depth of the pipe. It may

208

be necessary to strut flexible pipe inside to prevent distortion during this operation. The fill around the top half of the pipe, up to 300 mm above the crown should be compacted to 85 to 90% Proctor density.

For backfill under roads, the top layers may have to be compacted to 90-95% Proctor density, while for runways 110% may be required. 85% is normally sufficient in open country and densities as low as 80% are obtained if control is bad (Reitz, 1950).

To avoid damage to pipe or coatings, fill around the pipe should not have stones in it. Surplus spoil may be spread over the trench width to compensate for settlement or carted away. Reinstatement of topsoil and vegetation, provision of land drainage, and fencing off of working areas should be considered (see also Sowers, 1956).

THRUST BORES

To avoid excessively deep excavations, or to avoid disrupting traffic on a road over a pipe, the pipe may be jacked through the soil for short lengths. A hole is dug at each end of the length to be jacked. A jack is set up in one hole with its back against a thrust block or timber and the pipe set up on rails in front of the jack. Small pipes (less than 600 mm diameter) may have a sharp head or shoe fitted to the front to ease the jacking load and maintain the direction of the pipe, as the slightest obstacle will deviate the pipe. Occasionally the outside of the pipe is lubricated with bentonite to reduce friction. Pushing should never stop for long or the pipe may stick. With larger pipes the soil ahead of the pipe is dug out by auger or hand and removed through the pipe. The pipe is then pushed in further, the jack retracted and another pipe inserted behind the pipes in the bore and the process repeated. The jacking force required to push a pipe with a shoe may be up to 1 200 tons, but if an auger is used, a 100 ton jack may suffice.

PIPE BRIDGES

Pipes frequently have to span rivers or gorges. For short spans, rigid jointed pipes may have sufficient strength to support them-

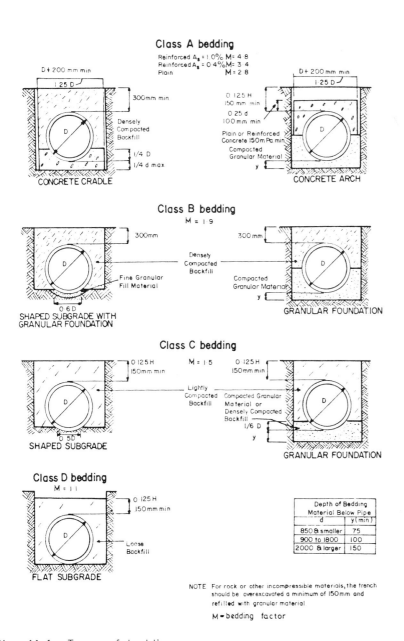

Fig. 11.1   Types of beddings

selves plus the fluid. For larger spans, it may be necessary to support the pipe on trusses or concrete bridges, or hang the pipe from an existing traffic bridge. If a truss bridge is to be constructed for the pipe, the pipe could act as the tension member at the bottom, or the compression member at the top. An attractive form of bridge is one with the pipe supported from suspension cables. Two cables are preferable to one, as they may be spaced apart and the hangers attached at an angle to the vertical plane through the pipe. This arrangement helps to support the pipe against wind forces and reduces wind vibrations. The pipe could also be designed to act as an arch, but again lateral support would be necessary – either two pipes could be designed to act together with cross bracing, or cable stays could be erected. In all cases except if supported on an independent bridge, the pipe should be rigid jointed; preferably welded steel. This means there will have to be some form of expansion joint along the pipe to prevent thermal stresses developing.

UNDERWATER PIPELINES

The laying of pipes underwater is very expensive, nevertheless it is often inescapable. In addition to river crossings, undersea laying is becoming common. Gas and oil lines from undersea beds or off-shore tanker berths are frequent. There are also many industries and towns which discharge effluent or sludge into the sea through undersea pipes. The method used for underwater laying will depend on conditions such as currents, wave height, type of bed and length of under-water pipe:

(1)   Floating and Sinking: In calm water the pipe may be jointed on shore, floated out and sunk. Lengths up to 500 m have been laid thus. Care is needed in sinking the pipe and winches on barges should be positioned along the line to assist.

(2)   Bottom Towing: Lengths of pipe up to a few kilometres may be assembled on land in strings, each up to 1 km long, and towed along the bed of the sea by a winch of a barge anchored at sea. Strings are jointed together on the shore. The pipe is buoyed up by air in the pipe or by buoys tied to the pipe, to reduce friction on the sea bed. If the pipe

is filled with air it may have to be weighted down by a concrete coating to get the correct buoyancy. Too light a pipe would be susceptible to side currents and may bend out of line. The stresses in the pipe due to the tension and bending caused by underwater currents or waves may be critical. The lateral drag per unit length due to waves and currents is

$$C_D w \, (u + v)^2 \, D/2g \qquad (11.1)$$

where C_D is the drag coefficient (about 0.9), w is the specific weight of the water, u is the wave velocity and v is the current velocity (or their components perpendicular to the pipe), D is the outside diameter of the pipe and g is gravitational acceleration.

(3)   Lay Barge: Pipes may be taken to sea in a pipe barge, jointed on a special lay-barge and lowered continuously down an arm (or stringer) into place. The barges proceed forward as laying proceeds.

(4)   Underwater Jointing: Underwater jointing is difficult as each pipe has to be accurately aligned. A diver will be required, which limits the method to shallow depths (Reynolds, 1970).

It is usually necessary to bury the pipe to avoid damage due to currents. Dredging is difficult as sediment may fall or be washed into the trench before the pipe is laid. The trench normally has to be over-excavated. A novel way of laying the pipe in soft beds is with a bed fluidizer (Schwartz, 1971). High-pressure jets of water fluidize the bed under the pipes and the pipe sinks into the bed. A number of passes of the fluidizer are required so that the pipe sinks gradually over a long length. Too deep a trench may cause bending stresses in the pipe at the head of the trench.

Most underwater pipes are steel with welded joints. Pipes and joints should be well coated and lined and cathodic protection is desirable under the sea.

JOINTS AND FLANGES

The type of joint to use on a pipeline will depend on the type of pipe, the facilities available on site, cost, watertightness of joint, strength and flexibility requirements. The following types of joints are used (EEUA, 1968) :-

(1)   Butt-welded: The ends of the steel pipe are trued and bevelled approximately 30° back, leaving a 2 mm root. The joint is clamped and/or tacked in places and then welded right around with one or more runs. Large pipes with thick walls may also require a weld pass inside, although it is difficult to make good the internal lining in pipes less than 600 mm bore.

(2)   Sleeve welded: One end of a steel pipe may be flared out to form a socket and the leading edge fillet welded to the barrel of the other pipe. If possible the joint should be welded inside too. (see Fig. 11.2). The external pipe coating and the internal lining, if possible, should be made good at the weld after all slag has been wire-brushed off.

Two methods of welding are available: Oxy-acetylene gas welding and metallic arc welding. Top-class welding is essential to achieve high weld strength and avoid cavities. A thorough testing programme should accompany the welding. Destructive testing should be confined where possible to the factory and to sample welds on short lengths of pipe but occasional field destructive tests may be necessary. Strips including the weld are cut from the pipe wall and bent over a predetermined radius. The weld is ground and etched to reveal air pockets or slag inclusions. Normally field tests will be non-destructive: either by X-ray, Gamma-ray, magnetic or ultra-sonic methods. X-rays are preferred for high definition and contrast but Gamma-ray equipment is more portable and useful for otherwise inaccessible corners. With large-bore pipes, the X-ray equipment may be inserted in the pipe and a film is wrapped around the pipe to receive the image from a single wall. For smaller pipes, double, overlapping expo-

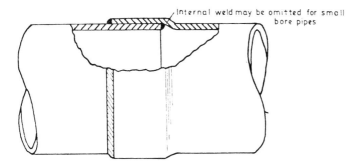

Fig. 11.2   Sleeve-welded joint

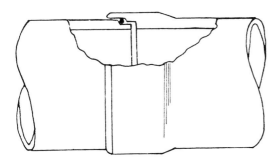

Fig. 11.3   Push-in type joint

sures are made from outside, thereby exaggerating defects in the near wall.

Welding of high tensile steels require better craftmanship that for low-tensile steels. Pre-heating of the pipe ends, and post-heating for stress relief are normally desirable.

(3)   Screwed: Threaded and screwed ends and sockets are only used on small steel pipes.

214

(4)   Spigot and socket: Concrete, cast iron and plastic pipes fre-
        quently have one end socketed and the other end plain. The
        male end has a rubber sealing gasket fitted over it and the
        socket is forced over the ring. (see Fig. 11.3). Various
        shaped rubber rings are used. Another way of sealing the joint
        is to caulk it with bitumen compound, cement mortar, lead or
        resin. Joints fitted with rubber rings are often caulked as well
        to hold the ring in place, improve sealing and prevent dirt
        getting into the joint. Spigot and socket joints with rubber
        rings can normally accommodate one or two degrees of
        deflection.

(5)   Clamp-on joints: Various proprietry joints are available for
        jointing plain ended pipes. They normally involve clamping
        a rubber seal between each pipe barrel and a cover sleeve.
        Fig. 11.4 illustrates one such type of joint (also referred
        to as a Viking Johnson or Dresser coupling). These joints
        accommodate movement easily, and may be designed to take
        several degrees of deflection between pipes or longitudinal
        movement. If the joint is to resist tension, tiebolts may be
        fixed to the barrel of each pipe by brackets to take the
        tension. The brackets cause local bending and tension stresses
        in the walls of the pipe and may cause damage unless care-
        fully designed. Normally four or even two tie bolts are
        sufficient. The diameter of the bolts is selected to take the
        tension and the brackets designed so that the bolts are as
        close as possible to the barrel to minimize eccentricity. The
        brackets are normally "U" shaped in plan with the legs
        welded parallel to the pipe axis. The length of the legs of
        the bracket should be such that longitudinal shear and bend-
        ing stresses are acceptable. In fact it is impossible to elim-
        inate all bending stresses on the pipe barrel with brackets
        and for this reason rings, like flanges, may be preferable
        for transferring the tie-bolt load to the pipe wall. The
        Victaulic coupling used for cast iron pipes, uses lips on the
        ends of the pipes to transfer tension to the coupling.

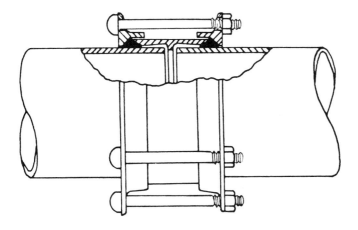

Fig. 11.4   Slip-on type coupling.

(6)   Flanged: Steel and cast iron pipes are often flanged at the
ends, especially if pipes or fittings are likely to be removed
frequently. Faces are machined and bolted together with 3
mm rubber or other insertion gasket between. Flanges are
drilled to standard patterns according to the diameter and
working pressure.

Full face gaskets, with holes for bolts and the bore, are
used for high-pressure joints and cast iron or soft metal
flanges. Joint rings with an outside diameter slightly less
than the inside diameter of the bolt holes are used for thick
steel flanges and low-pressure pipes. Some flanges have a
raised face inside the bolt circle for use with joint rings.
Flanges used with joint rings tend to dish when the bolts
are tightened though. The method of attaching the flanges
to the pipe varies. The simplest method is to slip the flange
over the pipe but to leave approximately 10 mm extending
beyond the end of the pipe. A fillet weld is then carried
around the front of the pipe and at the back of the flange
around the pipe barrel. Alternatively both or one of the
inside edges of the flange may be bevelled for welding.
A method preferred for high pressure work is to have a lip

on the flange projecting over the end of the pipe so that the bore at the joint will be flush after welding.

Jointing and thrust flanges are fairly thick, as indicated by the standard codes of practice, and must be securely welded to the pipe or cast integrally. Two other types of flanges are common:-

(1) Puddle flanges are used on pipes which are cast in walls of water retaining structures. The object of these flanges is to prevent leakage of water and they are not necessarily as thick as flanges which are designed to transfer longitudinal pipe stress, although the flange should be securely attached to the pipe wall.

(2) Blank flanges are, as the name implies, installed at blank ends. Their thickness should be calculated to resist bending moments at the edges and centre. For a circular disc rigidity clamped at the edges the maximum radial bending stress occurs at the perimeter bolts, and is equal to $3\ pd^2/16t^2$ where p is the fluid pressure, d is the pitch circle diameter and t is the flange thickness. If the edges were not rigidly clamped the maximum stress would occur at the centre and would be 50% in excess of this amount.

COATINGS

Buried steel pipes are subject to corrosion and damage unless the pipe is coated. Coatings should ideally be resistant to scratching during transport and laying of the pipe, to moisture, chemical and biological attack, electric currents and temperature variations. They should be hard enough to prevent damage during handling and due to stones in the trench, yet sufficiently adhesive to adhere well to the pipe wall and flexible enough to withstand the flexing of the pipe wall (PPI, 1975 ; Cates, 1953).

The most common coating for pipes is a thin adhesive coat followed by a coating reinforced with fibres and then possibly an outer wrapping. The pipe surface is initially cleaned by wire brushing, sandblasting or acid pickling, and the prime coat is then applied by spray, brush or dipping the pipe in a bath. Bitumen or coal tar enamel is the preferred prime coating. After the primary coat the pipe may be spirally wrapped with impregnated felt or woven glass fibre matting. This is sometimes followed by paper or asbestos felt impregnated with bitumen or coal tar. The pipe is then whitewashed to assist in detecting damage and to shield the coating from the sun. The coating may be applied in the field after welding the pipe joints, or in the factory, in which case the ends are left bare for jointing and coated in the field. The total thickness of coating should be at least 5 mm.

Other types of coatings include an asbestos fibre bitumen mastic 3 to 6 mm thick, coal tar pitch, epoxy paints, PVC or polythene tapes (either self adhesive or bedded on an adhesive), resins or plastics, cement mortar and zinc applied by galvanizing. Exposed pipes may be primed and painted with bitumen based aluminium or enamel (AWWA, 1962).

Cement mortar coatings offer additional resistance to buckling in the case of large bore thin walled pipes. Cement mortar coatings are usually 12 to 20 mm ($\frac{1}{2}$ to 3/4 inch) thick (AWWA, 1954, 1962).

Finished coatings may be checked for flaws, pin holes etc. by means of a Holiday detector. An electrical conductor in the form of metal brushes or rolling springs is run along the pipe coating. An electrical potential is applied across the coating and a current is observed when flaws in the coating are detected.

LININGS

Steel pipes are lined to resist internal corrosion and minimize the friction losses. Unlined steel pipe may be oxidized by corrosive substances in the fluid. In the case of solids-conveying systems in particular, the oxide is rapidly scraped off leading to further corrosion. Corrosion of water-conveying pipes may be inhibited by maintaining a high pH e.g. by adding lime. Lime on the other hand could cause carbonate scaling.

The most popular linings are bitumen (3 to 5 mm thick) or coal tar enamel (2 to 3 mm thick). Bitumen, which is a by-product of petroleum, is the cheaper of the two.

Before applying the lining the pipe wall is cleaned by sandblasting or other methods and the lining is then applied by brush, spray, dipping or spinning to obtain a smooth surface. Spun enamel in particular provides a smooth finish. Coal tar enamel is also more resistant to moisture than bitumen, although it is more brittle and consequently subject to damage by impact and flexing of the pipe. Plasticised coal tar enamels have, however, now been developed to overcome the problems of brittleness.

Epoxy paints are also used successfully for linings, although careful application is necessary to ensure successive coats adhere to each other. The recommended thickness of the lining varies with the type of paint, but it is normally of the order of 0.3 mm applied in 2 to 4 layers. Coal tar epoxy linings should be avoided for portable water pipes as they taint the water. Lead based primers should also be avoided as they are toxic.

Cement mortar lining, applied centrifugally by spinning the pipe, is also used for large bore pipes. The applied lining is usually 6 to 12 mm (1/4 to $\frac{1}{2}$ inch) thick. Mortar linings have been applied successfully to pipes in the field. The old surface is first thoroughly cleaned by wire-brushing or sand blasting, and then coated centrifugally by a machine drawn or propelled slowly along the line (Cole, 1956).

On pipes over about 600 mm bore, most factory linings may be made good at field joints manually or by mechanical applicators.

CATHODIC PROTECTION

Despite the use of protective coatings, corrosion of the walls of steel pipes often occurs through flaws, pin-holes or at exposed fittings. Corrosion is due primarily to two causes; galvanic corrosion and stray current electrolysis (Uhlig, 1948).

Galvanic Corrosion

When two dissimilar materials are connected through an electrolyte, current may flow from one material to the other. The resulting electric current flows from the anode to the cathode through the electrolyte. Particles or ions, leave the anode causing corrosion. The cathode is not attacked.

Such action may take place when two dissimilar metals are in contact in conductive soil (for instance fittings of a different material to the pipe). Another more frequent form of galvanic corrosion occurs with pipes in corrosive soils. The effect is particularly marked in soil with varying characteristics, differential oxygen concentrations or in water with high chemical content, especially sulphur. Corrosion is also caused by biochemical action in the soil which is a type of galvanic action resulting from bacteria. A method which has been used successfully to counteract soil corrosion was to add lime to the trench backfill. Stress concentration in the steel in an electrolyte may also lead to corrosion. The amount of corrosion due to the last named two influences is usually small.

Potential corrosive areas may be detected by soil resistance tests, pipe – soil potential tests or measurement of the current in the pipe.

If the soil is at all suspect, or as a standard practice for major pipelines, a soil resistivity survey should be conducted. Soil resistivity is measured in ohm – centimetres. Readings are normally taken in-situ using a bridge circuit or by measuring a current and the associated volt drop. Standard probes are available for these measurements, but any major surveys should be done by an experienced corrosion engineer (Schneider, 1952).

A highly conductive soil may have a resistivity of 500 ohm cm, and a poorly conductive soil, more than 10 000 ohm cm.

The potential difference between a buried pipe and the soil is important in evaluating corrosive conditions. The potential is measured by connecting a voltmeter between the pipe and a special electrode in contact with the soil. A copper sulphate half cell is frequently used as the electrode under normal conditions. The potential of a pipe is 0.5 to 0.7 volts below that of the surrounding soil. If the pipe is at a higher voltage than 0.85 volts below that of the soil, currents are likely to flow from pipe to soil, thereby corroding the pipe.

To prevent this corrosion, a sacrificial anode may be connected by a conductor to the pipe in the vicinity of possible corrosion (Fig. 11.5). The sacrificial anode is buried in a conductive surround, preferably below the water table and currents will tend to leave from the anode instead of the pipe, thereby limiting the corrosion to the anode. If the anode is sufficiently dissimilar from the steel pipe to cause galvanic action no external electrical potential need be applied. In fact, a resistor may sometimes be installed in the connection to limit the current. Common sacrificial anodes are magnesium, zinc and aluminium, or alloys of these metals. Magnesium has the greatest potential difference from iron, has a high electro-chemical equivalent (ampere hours per kilogram of material) and is fairly resistant to anodic polarization. (A layer of hydrogen ions may replace the ions leaving the anodes, thereby actively insulating the metal against further attack. This is termed polarization). Magnesium anodes give 200 to 1 200 ampere hours per kg. Magnesium anodes are normally designed for about a 10 year life but zinc anodes often last 20 or 30 years.

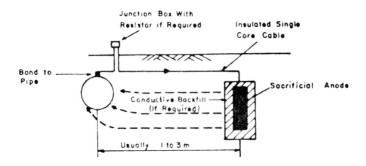

Fig. 11.5   Sacrificial anode installation for cathodic protection

The spacing and size of anodes will be determined by the current requirements to bring the pipe to a safe potential (at least 0.85 volts below the soil potential along the entire length of the pipe). The spacing of sacrificial anodes will vary from 3 m in poorly conductive soil and poorly coated pipe to 30 m in highly conductive soil provided the pipe is well coated. The most reliable way of estimating the required current is to actually drain current from the pipe by means of a ground bed and d.c. source and measure the resulting pipe to soil potential along the pipe.

As a rule of thumb, the current required to protect a pipe against corrosion is $i = 10/r$ per square metre of bared pipe surface, where i is the current in amps, r is the soil resistance in ohm cm and a factor of safety of 3 is incorporated. The area of exposed pipe may be as low as 0.5 per cent for a good coating, rising to 20 per cent for a poor coating.

Once the required draw-off current per km of pipe is known, the total anode size may be calculated. In highly conductive soils (less than 500 ohm cm) large anodes (20 kg) may be used but in soils with lower conductivities (greater than about 1 000 ohm cm) a number of smaller anodes should be used to ensure an adequate current output.

If an electricity supply is available, it is usually cheaper to use an impressed current type of protection (Fig. 11.6) instead of a sacrificial anode. A transformer-rectifier may be installed to provide the necessary dc current.

With impressed current installations, the anode need not be self-corrosive. Scrap iron, or graphite rods buried in coke fill, are frequently used. Steel anodes will quickly corrode in highly conductive soils and for this reason high silicon cast iron anodes are preferred.

The type of protection to use will depend on the long-term economics. Impressed current installations are frequently the cheapest in the long run, as maintainance costs are low and fewer installations are required than for sacrificial anodes. In the case of impressed current installations a relatively high voltage may be applied, which in turn protects a longer length of pipe than a sacrificial anode could. The pipe-soil potential falls off rapidly away from the applied voltage though, consequently a high voltage may have to be applied to protect a long length. The pipe-to-soil potential in the immediate vicinity of the point should not, in general exceed 3 volts (the pipe potential being below that of the soil). Larger voltages may damage coatings. Impressed current installations may be at 1 to 100 km spacing depending on the quality of the pipe coating. Impressed current installations are preferable to sacrificial anodes if the soil resistivity is higher than about 3 000 to 5 000 ohm cm.

222

Considerable economic savings are often achieved by protecting only "hot spots", or points subject to aggressive attack, and if occasional shutdowns can be tolerated this should be considered.

Stray Current Electrolysis

The most severe form of corrosion is often caused by stray dc currents leaving a pipe. Railways and other users of dc current return currents through rails, but if there is another conductor such as a steel pipe nearby, a proportion of the current may flow through the conductor instead of the rail. Where the current leaves the pipe, steel will be corroded at a rate of 9 kg (20 lbs) per year per ampere of current. The current may be detected by actual current measurements or from pipe/soil potential measurements. Continuous recordings should be taken over a day, as the currents may fluctuate with time.

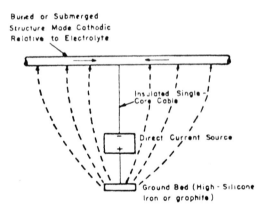

Fig. 11.6   Impressed current corrosion protection.

The corrosion associated with stray currents may be prevented by connecting the pipe at the point where the current leaves it, to the destination of the current, with a conductor. The current may leave the pipe along some length, in which case a current would have to be impressed on the pipe to maintain the voltage sufficiently low to prevent current escaping. A ground bed and transformer rectifier may be required for this protection.

When a pipe is cathodically protected, care should be taken that the mechanical joints are electrically bonded, by welding a cable across them if necessary. Branch pipes may have to be insulated from the main pipe to control the currents.

THERMAL INSULATION

Fluid in the pipeline may often be heated or cooled by the surroundings. Ice formation in arctic climates and heating in tropical climates are sometimes a serious problem. Heat is transferred to or from the interior of an exposed pipe in a number of ways: by conduction through the pipe wall and wrapping and a boundary layer of fluid inside the pipe wall, by radiation from the external face of the pipe and by convection and wind currents in the air surrounding the pipe.

TABLE 11.1   Thermal conductivities

| Material | Thermal conductivity | |
	$\dfrac{k\ cal}{m\ sec\ C°}$	$\dfrac{Btu\ in}{sq\ ft\ hr\ F°}$
Water	0.000 14	4
Air	0.000 005	0.15
Steel	0.014	420
Bituminized Wrapping	0.000 01	0.3
Concrete	0.0002	6
Slag wool	0.000 01	0.3

EEUA, 1968.

The heat loss by conduction through an homogeneous pipe wall is proportional to the temperature gradient across the wall and the heat transferred per unit area of pipe wall per unit time is

$$\frac{Q}{AT} = \frac{k}{t} \Delta\theta \qquad (11.2)$$

where Q is the amount of heat conducted in kilocalories or Btu, A is the area of pipe surface, T is the duration time, $\Delta\theta$ is the temperature difference across the wall, t is the wall thickness and k is the thermal conductivity, tabulated in Table 11.1 for various materials.

An equation was developed by Riddick et al (1950) for the total heat transfer through the wall of a pipe conveying water. The equation has been modified to agree with relationships indicated by EEUA (1964) and is converted to metric units here:

Rate of heat loss of water kcal/kg/sec

$$= \frac{(\theta_i - \theta_o)/250D}{\dfrac{t_1}{k_1} + \dfrac{t_2}{k_2} + \dfrac{1}{K_f} + \dfrac{1}{K_r + K_c}} \tag{11.3}$$

and since the specific heat of water is unity, the rate of drop in temperature in degrees Centigrade per sec equals the heat loss in kcal/kg/sec, where

θ_i	=	temperature of water C°
θ_o	=	ambient temperature outside C°
D	=	diameter m
t_1	=	thickness of pipe m
k_1	=	conductivity of pipe wall kcal/m sec C°
t_2	=	thickness of coating m
k_2	=	conductivity of coating kcal/m sec C°
K_f	=	heat loss through boundary layer = $0.34(1 + 0.020\theta_i)v^{0.8}/D^{0.2}$ kcal/m² sec C°
v	=	water velocity m/s
K_r	=	heat loss by radiation = $0.054\ E(\frac{\theta_o + 273}{100})^3\ 10^{-3}$ kcal/m² sec C°
E	=	emissivity factor = 0.9 for black asphalt and concrete 0.7 for cast iron and steel 0.4 for aluminium paint
K_c	=	heat loss by convection = $0.000\ 34\sqrt{1 + 0.78V}\ (\frac{\theta_i - \theta_o}{D})^{0.25}$ kcal/m² sec C°
V	=	wind velocity km/hr

The heat generation by fluid friction is usually negligible although it theoretically increases the temperature of the fluid.

Example

An uncoated 300 mm diameter pipeline 1 000 m long with 5 mm thick steel walls, exposed to a 10 km/hr wind, conveys 100 ℓ/s of water at an initial temperature of 10°C. The air temperature is 30°C. Determine the end temperature of the water.

$$E = 0.7$$

$$v = 0.1/0.785 \times 0.3^2 = 1.42 \text{ m/s}$$

$$K_f = 0.34 \times 1.42^{0.8}/0.3^{0.2} = 0.67$$

$$K_r = 0.054 \times 0.7 \ (\frac{30 + 273}{100})^3 \times 10^{-3} = 0.001$$

$$K_c = 0.34 \sqrt{1 + 0.78 \times 10} \ (\frac{30 - 10}{0.3})^{0.25} \times 10^{-3} = 0.0029$$

$$\frac{t_1}{k_1} = \frac{0.005}{0.014} = 0.35$$

$$\Delta\theta /\text{sec} = \frac{(30 - 10)/(250 \times 0.3)}{0.35 + 1/0.67 + (0.001 + 0.0029)} = 0.00103°C/\text{sec}$$

Temperature rise over 1 000 m = $\frac{1\ 000}{1.42} \times 0.00103 = 0.7°C$

The insulation of industrial pipework carrying high or low temperature fluids is a subject on its own. The cost of the heat transfer should be balanced against the cost of the lagging.

Heat transfer to or from buried pipelines depends on the temperature of the surroundings, which in turn is influenced by the heat transfer. The temperature gradient and consequently rate of heat transfer may vary with time and are difficult to evaluate. If the temperature of the surroundings is known, Equ. 11.3 may be used, omitting the term $1/(K_r + K_c)$.

REFERENCES

American Concrete Pipe Assn., 1970. Concrete Pipe Design Manual, Arlington.
AWWA Standard C602, 1954. Cement Mortar Lining of Water Pipelines in Place, N.Y.

226

AWWA Standard C203, 1962. Coal Tar Enamel Protective Coatings for Steel Water Pipe 30 Ins. and Over, N.Y.

AWWA Standard C205, 1962. Cement Mortar Protection Lining and Coating for Steel Water Pipe 30 Ins. and Over, N.Y.

BSCP 2010, 1970. Part 2, Design and Construction of Steel Pipelines in Land, BSI, London.

Cates, W.H., 1953. Coating for steel water pipe, J. Am. Water Works Assn. 45(2).

Cole, E.S., 1956. Design of steel pipe with cement coating and lining, J. Am. Water Works Assn., 48(2).

Concrete Pipe Assn., 1967. Bedding and Jointing of Flexibly Jointed Concrete Pipes, Techn. Bul. No. 10, Tonbridge.

Engineering Equipment User's Assn., 1964. Thermal Insulation of Pipes and Vessels, Handbook No. 12, Constable, London.

Engineering Equipment Users Assn., 1968. Pipe Jointing Methods, Handbook No. 23, Constable, London.

Pipes and Pipelines International, 1975. Pipeline protection review, 20(4).

Reitz, H.M. 1950. Soil mechanics and backfilling practice, J. Am. Water Works Assn., 42(12).

Reynolds, J.M., 1970. Submarine pipelines, Pipes and Manual. 3rd Ed. Scientific Surveys, London.

Riddick, T.M., Lindsay, N.L. and Tomassi, A., 1950. Freezing of water in exposed pipelines, J. Am. Water Works Assn., 42(11).

Schneider, W.R., 1952. Corrosion and cathodic protection of pipelines, J. Am. Water Works Assn., 44(5).

Schwartz, H.I., 1971. Hydraulic trenching of submarine pipelines, Proc. Am.Soc. Civil Engrs. 97(TE4)

Sowers, G.F., 1956. Trench excavation and backfilling, J. Am. Water Works Assn., 48(7).

Uhlig, H.H., 1948. The Corrosion Handbook, Wiley, N.Y.

LIST OF SYMBOLS

A	–	area
A_s	–	area of steel
B	–	width of trench
C_D	–	drag coefficient
d	–	inside diameter
D	–	outside diameter
E	–	thermal emissivity factor
f	–	stress
g	–	gravitational acceleration
H	–	backfill above top of pipe
i	–	current in amps

K_c	–	heat loss by convection
K_f	–	heat loss through water film
K_r	–	heat loss by radiation
k_1	–	conductivity of pipe wall
k_2	–	conductivity of coating
M	–	bedding factor
p	–	pressure
Q	–	amount of heat
r	–	resistance in ohms
T	–	time
t_1	–	thickness of pipe wall
t_2	–	thickness of coating
u	–	component of wave velocity perpendicular to pipe
v	–	mean water velocity, or component of current velocity perpendicular to pipe
V	–	wind velocity
w	–	specific weight of water
y	–	depth of bedding material below pipe
θ_i	–	temperature inside pipe
θ_o	–	temperature outside pipe

CHAPTER 12

PUMPING INSTALLATIONS

INFLUENCE OF PUMPS IN PIPELINE DESIGN

The pipeline engineer is frequently concerned with the design of pumping lines. That is, the fluid is forced through the pipe by a pump as opposed to a gravity fed line. The design of the pipe is inter-related with the pumping equipment selected and vice versa. Pumps must overcome line friction losses in addition to net static head, hence the pumping head and power requirement are a function of the pipe diameter. The pipe wall thickness is in turn based on internal pressures. Therefore friction head and fluid velocity (which affects water hammer head) also affect the wall thickness of the pipe.

The engineer therefore cannot design one without considering the other. The system must be designed in an integrated manner. Pipeline engineers should therefore be aware of the various types of pumps, their limitations, their characteristics and costs. Stepanoff (1957), IWE (1969) and Addison (1955), for example, give more complete information. The pump size and suction head requirements will also have to be considered in the pumpstation design. The design of the pump, i.e. the optimum casing or impeller shape and material, is more the concern of the manufacturer. In fact it is covered in books such as those by Karassik and Carter (1960), Kovats (1964) and KSB (1968). Similarly electrical power supply, motor design and control equipment are so specialized that they should be selected or designed by electrical engineers. It is from the pipeline engineer's point of view that the following information is presented.

TYPES OF PUMPS

Positive Displacement Types

Least used in pipeline practice nowadays, but still used in some industrial applications, are positive displacement pumps.

The reciprocating pump is one such type. A piston or ram is forced to move to and fro in a cylinder. One-way valves permit the liquid to enter on a reverse stroke and exit into the receiving conduit on a forward stroke of the piston.

The piston can be replaced by a flexible diaphragm. There are also rotary positive displacement pumps. These include spiral types and intermeshing gears.

Centrifugal Pumps

Rotodynamic pumps move liquid by imparting a velocity and hence pressure to it. This action may be in the form of an axial flow like through a propeller. Liquid is forced along a tabular casing by rotating blades. There may be guide vanes in the casing to reduce swirl. This type of pump is most common for large flows and low heads. For high heads the centrifugal pump is preferred. Here rotation of an impeller creates an outward (radial) centrifugal flow.

The latter form is the most common, and the centrifugal pump in various forms is universally accepted as practically standard in waterworks engineering. This type of pump is capable of handling flows up to 20 cumecs. For high flows and to permit flexibility, pumps in parallel are common. The head which can be generated by an impeller is limited to a little over 100 metres of water, but by assembling a number of impellers in series a wide range of heads is possible.

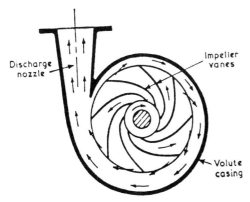

Fig. 12.1   Sectional elevation of volute-type centrifugal pump
(Webber, Fluid Mechanics for Civil Engineers, 1971)

230

The inlet condition must be controlled to avoid cavitation or air entrainment and the permissible maximum suction pressure is related to the delivery pressure of the first pumping stage. In some cases instead of connecting impellers in series on a common shaft a number of independent pump units are installed in series. This permits more flexibility particularly if one of the units is to be driven by a variable speed motor.

The driving device in a centrifugal pump is termed the impeller. It may be a series of vanes shaped to fling the water from a central eye or inlet towards the outer periphery. Here it is directed by a volute casing into the discharge nozzle. The efficiency of this type of pump may be as much as 90 percent although pump and motor units together rarely have efficiences above 80 percent except in large sizes. Pumping installations for unusual duties such as pumping solids in suspension can have efficiencies as low as 30 percent.

Usually the inlet to the impeller eye is on one side, and the driving shaft to the motor on the other side of the impeller. In some situations an inlet may be on each side in which case it is referred to as a double entry pump. This involves a more expensive casing but it results in a balanced thrust in the direction of the shaft.

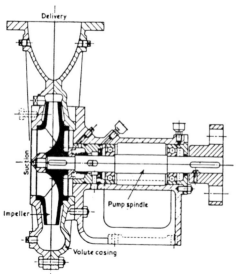

Fig. 12.2 Section through a single entry horizontal spindle centrifugal pump

Pump units may be mounted with the shaft vertical or horizontal. The horizontal shaft with a split casing is the most practical from the maintenance point of view. Where space is limited, or a deep suction well is required, a vertical spindle arrangement may be preferable. The motor is then mounted on a platform above the pump.

TERMS AND DEFINITIONS

Head

The head in its general sense is the energy per unit weight of water. It comprises elevation above a certain datum, z, plus pressure head p/w plus velocity head $v^2/2g$.

Total Head

Total head usually refers to the excess of discharge head to inlet head and it is the head generated by the pumps (see Fig. 12.3).

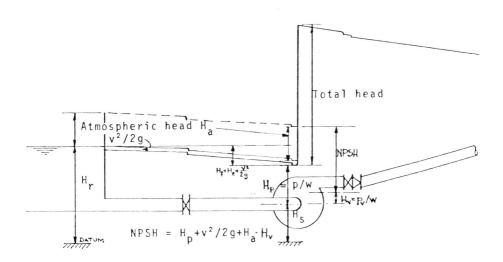

Fig. 12.3 Definition of net positive suction head.

Net Positive Suction Head

In order to avoid cavitation, air entrainment and a drop-off in pumping efficiency a pump needs a certain minimum suction head. The net positive suction head (NPSH) required can usually be indicated by a pump manufacturer as a function of rate of pumping. The requirement is referred to as the net positive suction head. It is generally expressed as the absolute head above vapour pressure required at the pump inlet. It is calculated as follows:

$$\text{NPSH} = H_p - p_v/w \tag{12.1}$$

$$= H_r + H_a - H_s - H_f - H_e - p_v/w \tag{12.2}$$

where H_p = head at inlet, relative to centre line of inlet, and equal to absolute pressure head plus velocity head.

p_v = vapour pressure

w = unit weight of water

H_r = suction reservoir head above datum

H_a = atmospheric pressure head (about 10 m of water at sea level)

H_s = elevation of centreline of pump suction above datum

H_f = friction head loss between reservoir and pump

H_e = turbulence head loss between reservoir and pump

NPSH requirement is a subjective figure as there is not a sharp drop-off in efficiency or increase in noise level at this value. It increases with discharge. It is preferable to be on the conservative side when considering NPSH. The NPSH requirement is of the order of 10 percent of the head of the first stage of the pump but a more reliable figure should be obtained from the manufacturer's tests.

NPSH can be assessed by observing cavitation or from performance tests (Grist, 1974). As the NPSH is reduced cavitation commences, firstly at flowrates away from best efficiency flowrate but eventually at all flowrates. At an NPSH just less than that associated with maximum erosion due to cavitation the head generated by the impeller starts to decrease. The head rapidly decreases with decrease in NPSH. A limit of 3% decrease in head is recommended for establishing the NPSH. The NPSH required to avoid cavitation should be at least 3 times the 3% head drop NPSH, though. For this reason some degree of cavitation is normally acceptable.

Specific Speed

The specific speed N_s of a pump is a useful indication of the pump's capabilities. There are two slightly different definitions of specific speed and the more general one is given first:

Specific speed is a characteristic velocity of the rotating element relative to a characteristic velocity of the water, i.e. it is the velocity of the impeller relative to that of water.

The speed of the periphery of the impeller is proportional to ND where N is the impeller rotational speed and D is its diameter (units will be introduced later). From Bernoulli's theorum the water velocity is proportional to $\sqrt{gH}$ where H is the total head. Therefore N_s is proportional to ND/$\sqrt{gH}$. But discharge Q is proportional to vDb where b is the width of the impeller. For any given geometric proportion b is proportional to D, so

$$Q \propto \sqrt{gH}\ D^2 \tag{12.3}$$

Eliminating D one arrives at

$$N_s \propto NQ^{\frac{1}{2}} / (gH)^{3/4} \tag{12.4}$$

This is a dimensionless form of specific speed provided the units are consistent. Unfortunately most catalogues and text books omit the g term and express N in revolutions per minute (rpm) not radians per second. One encounters the expression

$$N_s = NQ^{\frac{1}{2}}/H^{3/4} \tag{12.5}$$

In imperial units Q is in gallons per minute and H in feet. In S.I. units Q is in cumecs (cubic metres per second) and H in metres.

The alternative definition of N_s is the speed in rpm of a geometrically similar pump of such size that it would deliver one cumec (or gal. per min) against 1 metre (or 1 foot) head. Then by substitution one arrives directly at the latter dimensionally dependent expression for N_s.

By reviewing the former definition of N_s, one obtains an idea of the size of impeller relative to pumping head. In fact a high N_s is associated with a low head and fast rotational speeds, whereas a low N_s implies a high head and low speed relative to that of water (water thus has a high radial component relative to tangential component).

Fig. 12.4 indicates the relationship between head per stage and specific speed for a number of pumps selected from real applications. N is in rpm and ω in radians/sec (ω = 2 π N/60).

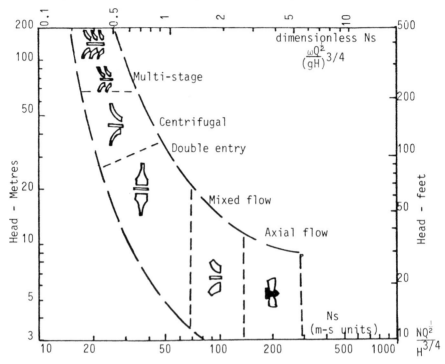

Fig. 12.4   Relationship between head and specific speed for pumps

IMPELLER DYNAMICS

The shape of the impeller blades in a centrifugal pump influences the radial flow pattern which in turn gives the pump certain characteristics. A basic understanding of relative velocities inside the pump is therefore useful in selecting a type of pump for a particular duty.

Referring to Fig. 12.5 the circumferential speed of the impeller is u, the velocity of the water relative to the impeller is w and the absolute water velocity is v.

Torque T on the water is force × radius

But force = change in momentum

Therefore imparted T = change in moment of momentum

$$= \Delta(mvr) \tag{12.6}$$

$$= \int r_2 v_2 \cos \alpha_2 \, dm - \int r_1 v_1 \cos \alpha_1 \, dm \tag{12.7}$$

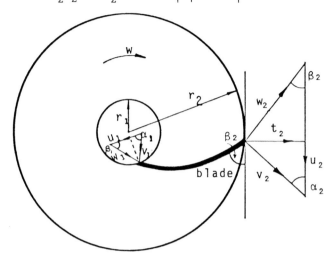

Fig. 12.5   Velocity vectors at impeller of centrifugal pump.

where dm is an element of mass discharge per unit time, and the integral is over all the vanes.

Power $P = T \omega$ where ω is the rotational speed $\tag{12.8}$

$\therefore P = \Delta(mvr)\omega$ $\tag{12.9}$

Now $r_1 \omega = u_1$ and $r_2 \omega = u_2$ $\tag{12.10}$

$\therefore P = \int u_2 v_2 \cos \alpha_2 dm - \int u_1 v_1 \cos \alpha_1 dm$ $\tag{12.11}$

Also $P = \rho g H Q$ where Q is the discharge rate $\tag{12.12}$

$\therefore P = gH \int dm$ since $\rho Q = \int dm$ $\tag{12.13}$

$\therefore$ Head $H = (u_2 v_2 \cos \alpha_2 - u_1 v_1 \cos \alpha_1)/g$ $\tag{12.14}$

The term $u_1 v_1 \cos \alpha_1$ is normally relatively small.

Now $v_2 \cos \alpha_2 = u_2 - t_2 \cot \beta_2$ $\tag{12.15}$

where t_2 is the radial component of v_2

$\therefore H \doteqdot (u_2 - t_2 \cot \beta_2)/g$ $\tag{12.16}$

and $Q = \pi D b t_2$ $\tag{12.17}$

$\therefore H \doteqdot A - BQ \cot \beta_2$ $\tag{12.18}$

where A and B are constants, g is gravity and ρ is water mass density. The blade angle β_2 therefore affects the shape of the pump H-Q characteristic curve, as illustrated in Fig. 12.6.

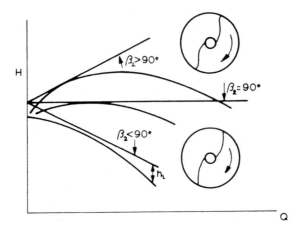

Fig. 12.6   Effect of blade angle on pump characteristics

PUMP CHARACTERISTIC CURVES

The head-discharge characteristic of pumps is one of the curves usually produced by the manufacturer and used to design a viable pumping system. This relationship is useful for pipe size selection and selection of combinations of pumps in parallel as will be discussed later.

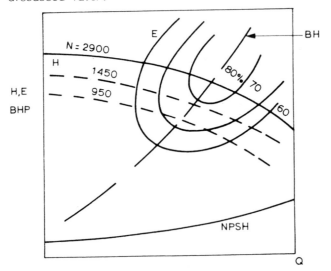

Fig. 12.7   Pump characteristic curves.

Break horse power (BHP) or shaft power required by the pump
and pump efficiency E are often indicated on the same diagram (see
Fig. 12.7). The duty H-Q curve may alter with the drive speed and
could also be changed on any pump by fitting different impeller
diameters. The BHP and E curves would also change then. The resultant head is proportional to N^2 and D^2, whereas the discharge is
proportional to N and D.

The required duty of a pump is most easily determined graphically. The pipeline characterisitcs are plotted on a head – discharge
graph. Thus Fig. 12.8 illustrates the head (static plus friction)
required for different Q's, assuming (a) a high suction sump water
level and a new, smooth pipe, and (b) a low suction sump level
and a more pessimistic friction factor applicable to an older pipe.
A line somewhere between could be selected for the duty line. The
effect of paralleling two or more pumps can also be observed from
such a graph.

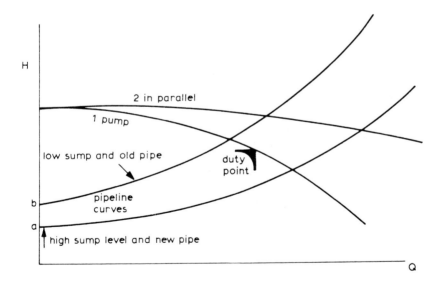

Fig. 12.8   Pipeline and pump characteristics.

Pumps in series are more difficult to control. Frequently, however, a booster pump station (Fig. 12.9) is required along a pipeline to increase capacity or to enable the first section of pipeline to operate within a lower design pressure. In such cases the pump curves are added one above the other, whereas for pumps in parallel, the discharge at any head is added, i.e. the abscissa is multiplied by the number of pumps in operation.

Booster pumps may operate in-line, or with a break pressure reservoir or sump on the suction side of the booster. In the latter case a certain balancing volume is required in case of a trip at one station. The hydraulic grade line must also be drawn down to reservoir level. In the former case water hammer due to a power failure is more difficult to predict and allow for.

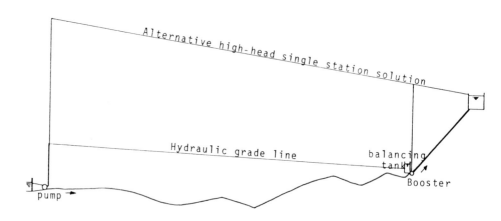

Fig. 12.9   Location of booster pump station.

MOTORS

The driving power of a pump could be a steam turbine, water wheel or most commonly nowadays, an electric motor. The rotational speed of the pump is controlled by the frequency of alternating current. The pump speed is approximately N = 60Hz divided by the number of pairs of poles, where N is the pump speed in revolutions per minute and Hz is the electrical frequency in cycles per second or Hertz. There is a correction to be made for slip between the rotor and starter. Slip can be up to 2 or 5 percent (for smaller motors).

Three types of AC motors are commonly used for driving pumps:
a) squirrel cage induction motors (asynchronous)
b) wound rotor (slip-ring) induction motors (asynchronous)
c) synchronous induction motors.

Squirrel cage motors are simple in design, robust, economic and require minimum maintenance. They are generally more efficient than the other two types. Their disadvantage is that they require large starting currents. However if this can be tolerated, a squirrel cage motor with direct-on-line starter offers the most economical and reliable combination possible. If direct-on-line starting is not permitted, it is usual to employ either star-delta or auto-transformer starters. Alternative starters compare as follows:-

Type of Starter	Starting Current (proportion of full load)	Starting Torque (full load = 1)
Direct - on	4.5 to 7.0	0.5 to 2.5
Star - Delta	1.2 to 2.0	0.3 to 0.8
Auto - transformer	as required	as required

Slip-ring induction motors are started by means of resistance starters and have the advantage that they can be smoothly brought up to full speed with relatively low starting current. They are normally used when squirrel cage motors are impermissible.

Synchronous motors are used on large installations. They have the advantage of a high load factor. They may be either induction motors or salient pole motors. They run at constant synchronous speed. Starting can be by direct-on-line, auto-transformer, resistance or reactor starters depending mainly on the permitted starting current.

Variable speed motors are more expensive and less efficient than constant speed motors. Belt drives on smaller units are possible although commutator motors, or motor resistances on slip-ring motors are used as well. Thyristor starters and variable speeds are now being used with savings in cost for small motors. Pole changing can also be used for selecting two speeds for squirrel cage motors.

PUMPSTATIONS

The cost of the structures to house pumps may exceed the cost of the machinery. While the design details cannot be covered here the pipeline engineer will need to concern himself with the layout as he will have to install the suction and delivery of pipework to each pump with control valves. The capacity of the suction sumps will effect the operation of the system, and there may be water hammer protection incorporated in the station. A more complete discussion of pumpstation layouts is given by Twort et al (1974).

The design of the sump and pump inlet pipework has an important bearing on the capacity of a pipeline. Head losses into the pump can effect its operational efficiency. Air drawn in by vortices or turbulence could reduce the capacity of the pipeline considerably. The awareness of the requirements of the entire pumping system therefore should be the duty of the pipeline engineer.

REFERENCES

Addison, H., 1955. Centrifugal and Other Rotodynamic Pumps, 2nd Ed., Chapman and Hall, London, pp 530.
Grist, E., 1974. Nett positive suction head requirements for avoidance of unacceptable cavitation erosion in centrifugal pumps, Proc. Cavitation conference, Instn. Mech. Engrs. London, p 153-162.
Institution of Water Engineers, 1969. Manual of British Water Engineering Practice. Vol.II, Engineering Practice, 4th ed. London.
Karassik, I.J. and Carter, R., 1960. Centrifugal Pumps, F.W. Dodge, N.Y. pp 488.

Kovats, A., 1964. Design and Performance of Centrifugal and Axial
    Flow Pumps and Compressors, Pergamon Press, Oxford, pp 468.
KSB, Pump Handbook, 1968. Klein, Schanzlin & Becker, Frankenthal,
    pp 183.
Stepanoff, A.J., 1957. Centrifugal and Axial Flow Pumps, Theory,
    Design and Application, 2nd Ed., Wiley, N.Y., pp 462.
Twort, A.C., Hoather, R.C. and Law, F.M., 1974. Water Supply,
    Edward Arnold, London, pp 478.
Webber, N.B., 1971. Fluid Mechanics for Civil Engineers, Chapman
    and Hall, London, pp 340.

LIST OF SYMBOLS

A,B   –   constants

BHP   –   break horsepower

b   –   width

D   –   diameter

E   –   efficiency

g   –   gravitational acceleration

H   –   head (subscript a – atmospheric, e – turbulence,

                  f – friction, p – inlet and s – elevation).

m   –   mass per unit time

N   –   pump rotational speed

NPSH   –   net positive suction head

N_s –   specific speed

p   –   pressure

P   –   power

Q   –   discharge rate

r   –   radius

T   –   torque

t   –   radial component of velocity

u   –   peripheral velocity

v   –   velocity

w   –   unit weight of water ρg. (9800 Newtons per cubic metre)

ρ –   unit mass of water

α, β –   angles

GENERAL REFERENCES AND STANDARDS

BRITISH STANDARDS

Steel Pipes

BS	534	Steel Pipes, Fittings and Specials
	778	Steel Pipes and Joints
	1387	Steel tubes and tubulars for screwing
	1965Pt1	But-welding pipe fittings. Carbon steel
	2633	Class 1 arc welding of ferritic steel pipes
	2910	Radiographic ex. welded circ. butt joints in steel pipe
	3601	Steel pipes and tubes – carbon steel – ordinary duties
	3602	Steel pipes and tubes for pressure purposes Carbon steel : high duties
	3603	Steel pipes and tubes for pressure purposes Carbon and alloy steel pipes
	3604	Low and medium alloy steel pipes
	3605	Steel pipes and tubes for pressure purposes Austenitic stainless steel

Cast Iron Pipe

	78-1	C I spigot and socket pipes – pipes
	78-2	C I spigot and socket pipes – fittings
	143	Malleable C I screwed pipes
	416	C I spigot and socket soil, waste and vent pipes
	437	C I spigot and socket drain pipes and fittings
	1130	C I drain fittings – spigot and socket
	1211	Cast (spun) iron pressure pipes
	1256	Malleable C I screwed pipes
	2035	C I flanged pipes and fittings
	4622	Grey iron pipes and fittings

Wrought Iron Pipe

	788	Wrought iron tubes and tubulars

1740   Wrought steel pipe fittings

Ductile Iron Pipe
    4772   Ductile iron pipes and fittings

Asbestos Cement Pipe
BS   486   A C pressure pipes
    582   A C soil, waste and ventilating pipes and fittings
    2010Pt4   Design and construction of A C pressure pipes in land
    3656   A C sewer pipes and fittings

Concrete
    556   Concrete pipes and fittings
    4101   Concrete unreinforced tubes and fittings
    4625   Prestressed concrete pipes (inc. fittings)

Clay Pipe
    65   Clay drain and sewer pipes
    539   Clay drain and sewer pipes – fittings
    540   Clay drain and sewer pipes
    1143   Salt glazed ware pipes with chemically resistant
              properties
    1196   Clayware field drain pipes

Plastic and Other Pipe
    1972   Polythene pipe (type 32) for cold water services
    1973   Spec. polythene pipe (type 425) for general purposes
    3284   Polythene pipe (type 50) for cold water services
    3505   UPVC pipes (type 1420) for cold water supply
    3506   UPVC pipe for industrial purposes
    3796   Polythene pipe (type 50)
    3867   Dims. of pipes O.D.
    4346   Joints and fittings for UPVC pressure pipe
    4514   UPVC soil and ventilating pipe
    4660   UPVC underground and drain pipe

4728	Resistance to constant internal pressure of thermoplastic pipe
2760	Pitch fibre pipes and couplings
CP312	Plastic pipework

Insulation

1334	Thermal insulation
4508	Thermally insulated underground piping systems
CP3009	Thermally insulated underground piping systems

Valves

BS 1010	Taps and valves for water
1212Pt2	Ball valves – diaphragm type
1218 &	
5163(m)	Sluice valves
1415	Valves for domestic purposes
1952 &	
5154(m)	Copper alloy gate valves
1953	Copper alloy check valves
2060	Copper alloy stop valves
2591	Glossary of valves
3464 &	
5150(m)	C I wedge and double disc valves
3948 &	
5151(m)	C I parallel slide valves
3952 &	
5155(m)	C I butterfly valves
3961 &	
5152(m)	C I stop and check valves
4090 &	
5153(m)	C I check valves
4133 &	
5157(m)	Flanged steel parallel slide valves
4312	Stop and check valves
5156(m)	Screwdown diaphragm valves
5158(m)	Plug valves
5159(m)	Ball valves

Jointing

10	Flanges and bolting for pipes etc.
21	Threads
1737	Jointing materials and compounds
1821	Oxy-acetylene welding of steel pipelines
1965	Butt welding
2494	Rubber joint rings
3063	Dimensions of gaskets
4504	Flanges and bolting for pipes etc.

Miscellaneous

1042	Flow measurement
1306	Non-ferrous pipes for steam
1553	Graphical symbols
1710	Identification of pipelines
2051	Tube and pipe fittings
2917	Graphical symbols
3889	Non destructive testing of pipes and tubes
3974	Pipe supports
4740	Control valve capacity

SOUTH AFRICAN BUREAU OF STANDARDS

Steel Pipe

SABS 62	Steel pipes and fittings up to 150 mm
719	Electrical welded low carbon steel pipes
720	Coated and lined mild steel pipes

Cast Iron Pipe

509	Malleable C I pipe fittings
746	C I soil, waste, water and vent pipes
815	Shouldered end pipes, fittings and joints

Asbestos Cement Pipe

546	C I fittings for A C pressure pipes
721	A C soil, waste and vent pipes and fittings

946	A C pressure pipe – constant internal diameter type
286	A C pressure pipes – constant outside diameter type
819	A C sewer pipes

Concrete Pipe

676	R C pressure pipes
677	Concrete non-pressure pipes
975	Prestressed concrete pipes
902	Structural design and installation of precast concrete pipelines

Glazed Earthenware Pipe

| 559 | Glazed earthenware drain and sewer pipes and fittings |

Plastic Pipe

SABS 791	UPVC sewer and drain pipe fittings
921	Pitch impregnated fibre pipes
966	UPVC pressure pipe
967	UPVC soil, waste and vent pipes
997	UPVC pressure pipes for irrigation
0112	Installation of PE and UPVC pipes
533	Black polyethylene pipes

Valves

144	C I single door reflux valves
191	Cast steel gate valves
664	C I gate valves

AMERICAN WATER WORKS ASSOCIATION

AWWAC201	Fabricating electrically welded steel water pipe
C202	Mill type steel water pipe
C203	Coal-tar enamel protective coatings for steel water pipe
C205	Cement mortar protective coatings for steel water pipe of sizes 30" and over

C206	Field welding of steel water pipe joints
C207	Steel pipe flanges
C208	Dimensions for steel water pipe fittings
C300	Reinforced concrete water pipe – Steel cylinder type – not prestressed
C301	Reinforced concrete water pipe – Steel cylinder type – prestressed
C302	Reinforced concrete water pipe – Non-cylinder type – not prestressed
C600 – 54T	Installation of C I watermains
C602	Cement mortar lining of water pipelines in place (16" and over)

AMERICAN PETROLEUM INSTITUTE

API Std.5A	Spec. for Casing, Tubing and Drill Pipe
Std.5AC	Spec. for Grade C – 75 and C95 Casing and Tubing
Std.5AX	Spec. for High Strength Casing and Tubing
Std.5L	Spec. for Line Pipe
Std.5LA	Spec. for Schedule 5 Alum. Alloy Line Pipe
Std.5LP	Spec. for Thermoplastic Line Pipe
Std.5LR	Spec. for Glass Fibre Reinforced Thermosetting Resin Line Pipe
Std.5LS	Spec. for Spiral Weld Line Pipe
Std.5LX	Spec for High Test Line Pipe
RP5CI	Care and Use of Casing, Tubing and Drill Pipe
RP5LI	Railroad transport of Line Pipe
RP5L2	Internal Coating of Line Pipe for Gas Transmission
RP5L3	Conducting Drop Weight Tear Tests on Line Pipe
Bul.5C2	Performance Properties of Casing, Tubing and Drill Pipe
Bul.5T1	Non-destructive Testing Terminology

AMERICAN SOCIETY FOR TESTING MATERIALS

Concrete Pipes

ASTM C14	Concrete Sewer, Storm Drain and Culvert Pipe
C76	Reinforced Concrete Culvert, Storm Drain and Sewer Pipe
C118	Concrete Pipe for Irrigation or Drainage
C361	Reinforced Concrete Low-Head Pressure Pipe
C412	Concrete Drain Tile
C443	Joints for Circular Concrete Sewer and Culvert Pipe with Rubber gaskets
C444	Perforated Concrete Pipe
C497	Determining Physical Properties of Concrete Pipe or Tile
C505	Non-reinforced Concrete Irrigation Pipe and Rubber Gasket Joints
C506	Reinforced Concrete Arch Culvert, Storm Drain and Sewer Pipe
C507	Reinforced Concrete Elliptical Culvert, Storm Drain and Sewer Pipe
C655	Reinforced Concrete D-load Culvert, Storm Drain and Sewer Pipe

Steel Pipes

ASA B36.10	Steel pipes

Cast Iron Pipes

A142	C I Pipes
A377	C I pipes
A121.1	C I pipes

Asbestos Cement Pipes

C296	A C Pressure Pipes
C500	A C Pressure Pipes
C428	A C Pipes and Fittings for Sewerage and Drainage

Plastic Pipes

    D2241   Unplasticised PVC pipes

USDI BUREAU OF RECLAMATION

    Standard Spec. for Reinforced Concrete Pressure Pipe

BOOKS FOR FURTHER READING

Addison, H., 1964. A treatise on Applied Hydraulics, Chapman Hall, London

Albertson, M.L., Barton, J.R. and Simons, D.B., 1960. Fluid Mechanics for Engineers, Prentice-Hall, 1960.

American Concrete Pipe Association, 1970. Design Manual-Concrete Pipe, Arlington.

Am.Soc.Civil Engs. and Water Polln. Control Federation, 1970. Design and Construction of Sanitary and Storm Sewers, Manual 37, N.Y.

Am.Water Works Assn., 1964. Steel Pipe - Design and Installation, Manual M11, N.Y.

Bell, H.S., 1963. (Ed.), Petroleum Transportation Handbook, McGraw Hill, N.Y.

Benedict, R.P., 1977. Fundamentals of Pipe Flow, Wiley Interscience, 531 pp.

Bureau of Public Roads, 1963. Reinforced Concrete Pipe Culverts - Criteria for Structural Design and Installation, U.S. Govt. Printing Office, Washington, DC.

Clarke, N.W.B., 1968. Buried Pipelines - A Manual of Structural Design and Installation, MacLaren and Sons, London.

Colorado State Univ., 1971. Control of Flow in Closed Conduits, Proc. Inst., Fort Collins.

Crocker, S. and King, R.C., 1967. Piping Handbook, 5th Ed., McGraw Hill.

Davis, C.V. and Sorensen, K.E., 1969. Handbook of Applied Hydraulics, 3rd Ed., McGraw Hill, N.Y.

Holmes, E., 1973. Handbook of Industrial Pipework Engineering, McGraw Hill, N.Y.

Instn. Water Engs., 1969. Manual of British Water Engg. Practice, 4th Ed., London.

Littleton, C.T., 1962. Industrial Piping, McGraw Hill, 349 pp

Martin, W.L., 1961. Handbook of Industrial Pipework, Pitman, London.

Nolte, C.B., 1978. Optimum Pipe Size Selection, Trans Tech Publications, 297 pp.

Rouse, H., 1961. Engineering Hydraulics, Wiley, N.Y.

Stephenson, D., 1984. Pipeflow Analysis, Elsevier, 204 pp.

Streeter, V.L., 1961. (Ed), Handbook of Fluid Dynamics, McGraw Hill, N.Y.

Twort, A.C., Hoather, R.C. and Law, F.M., 1974. Water Supply, 2nd Ed., A. Arnold, London.

Walski, T.M., 1984. Analysis of Water Distribution Systems. Van Nostrand Reinhold, N.Y. 275 pp.

Walton, J.H., 1970. Structural Design of Vitrified Clay Pipes, Clay Pipe Development Association, London.

Watters, G.Z., 1984. Analysis and control of unsteady flow in pipelines. 2nd Ed. Butterworths.

Young, D.C. and Trott, J.J., 1984. Buried Rigid Pipes. Elsevier Applied Science, London, 230 pp.

APPENDIX

SYMBOLS FOR PIPE FITTINGS

<u>GENERAL</u>

ELBOW		BEND
SLEEVED		JACKETED
TEE		FRONT VIEW OF TEE
FILLET WELDED TEE		BACK VIEW OF TEE
CROSS OVER		HANGER
BELLMOUTH		SIMPLE SUPPORT
TAPER	300 150	CHANGE IN DIA 100 75

<u>JOINTS</u>

FLANGED		ELECTRICALLY BONDED
ELECTRICALLY INSULATED		BUTT WELD
FLEXIBLE		SCREWED
SWIVEL		SOCKET
EXPANSION		SPIGOT & SOCKET
SLEEVE COUPLING		END CAP

<u>VALVES</u>

ISOLATING		BUTTERFLY
WEDGE GATE		NEEDLE
REFLUX	or	GLOBE
ROTARY PLUG		DIAPHRAGM
AIR		RELIEF

<u>MISCELLANEOUS</u>

HYDRANT		STRAINER
FLOW INDICATOR		SPRAY
SURFACE BOX		VENT
DRAIN		PLATE BLIND
MOTOR		HANDWHEEL

PROPERTIES OF PIPE SHAPES

Circle

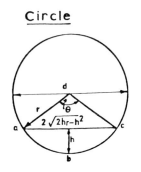

AREA $= \pi d^2/4$

CIRCUMFERENCE $= \pi d$

LENGTH OF ARC $abc = d.$ arccos $(1 - 2\frac{h}{d})$

AREA OF SEGMENT $abc = \frac{d^2}{4}$ arccos$(1 - 2\frac{h}{d}) - (d - 2h)\sqrt{hd - h^2}$

MOMENT OF INERTIA ABOUT DIAMETER $= \pi d^4/64$

MOMENT OF INERTIA ABOUT CENTRE $= \pi d^4/32$

Ring

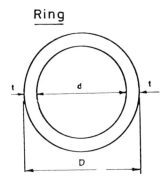

AREA $= \frac{\pi}{4}(D^2 - d^2) \approx \pi D t$ for small t

MOMENT OF INERTIA ABOUT DIAMETER $= \frac{\pi}{64}(D^4 - d^4) \approx \frac{\pi}{8}D^3 t$

MOMENT OF INERTIA ABOUT CENTRE $\frac{\pi}{32}(D^4 - d^4) \approx \frac{\pi}{4}D^3 t$

Ellipse

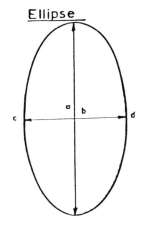

AREA $= \pi\frac{ab}{4}$

MOMENT OF INERTIA ABOUT AXIS $cd = \pi \frac{a^3 b}{64}$

MOMENT OF INERTIA ABOUT CENTRE $= \pi\frac{ab}{64}(a^2 + b^2)$

PROPERTIES OF WATER

Temperature		Specific Mass		Kinematic Viscosity		Bulk Modulus		Vapour Pressure	
C°	F°	kg/m³	lb/cu ft	m²/s	sq ft/sec	N/mm²	psi	N/mm²	psi
0	32	1 000	62.4	1.79×10^{-6}	1.93×10^{-5}	2 000	290 000	0.6×10^{-3}	0.09
10	50	1 000	62.4	1.31×10^{-6}	1.41×10^{-5}	2 070	300 000	1.2×10^{-3}	0.18
20	68	999	62.3	1.01×10^{-6}	1.09×10^{-5}	2 200	318 000	2.3×10^{-3}	0.34
30	86	997	62.2	0.81×10^{-6}	0.87×10^{-5}	2 240	325 000	4.3×10^{-3}	0.62

PROPERTIES OF PIPE MATERIALS

	Coef. of exp.per °C	Density kg/m³	Modulus of elasticity N/mm²	psi	Yield Stress N/mm²	psi	Tensile Strength N/mm²	psi	Poisson's ratio
Clay	5×10^{6}								
Concrete	10×10^{-6}	2 600	14 000–40 000	2×10^{6}–6×10^{6}	−70	−10 000	−2.1	−300	0.2
Asbestos cement	8.5×10^{-6}	2 500	24 000	3.5×10^{6}	17	2 500			0.25
Cast Iron	8.5×10^{-6}	7 800	100 000	15×10^{6}	150	22 000	225	23 000	0.3
Mild Steel	11.9×10^{-6}	7 850	210 000	31×10^{6}	210	30 000	330	48 000	
High Tensile Steel	11.9×10^{-6}	7 850	210 000	31×10^{6}	1 650	240 000	1 730	250 000	
Pitch Fibre	40×10^{-6}								
Polyethylene	160×10^{-6}	900	138	20 000	3	400	14	2 000	
Polyethylene (high density)	200×10^{-6}	955	240	35 000	22	3 200	32	4 600	0.38
PVC	50×10^{-6}	1 300			8	1 100	17	2 500	
UPVC	50×10^{-6}	1 400	3 500	0.5×10^{6}	40	+6 000	52	7 500	
Polypropylene	180×10^{-6}	912	150	22 000	25	3 600	30	4 500	

CONVERSION FACTORS

Length	1 inch = 25.4 mm
	1 ft = 0.3048 m
	1 mile = 1.61 km
Area	1 sq inch = 644 mm²
	1 acre = 2.47 ha
	1 sq ft = 0.0929 m²
	1 ha = 10^4 m²
Volume	35.31 cu ft = 1 m³
	1 gal.(imperial) = 4.54 litres
	1 US gal. = 3.79 litres
	1 barrel = 42 US gal. = 35 imp.gal.
	= 159 litres
Speed	1 ft/sec = 0.3048 m/s
	1 mph = 1.61 km/hr.
Acceleration	32.2 ft/sec² = 0.981 m/s² = g
Discharge	1 mgd (imperial) = 1.86 cusec
	35.3 cusec = 1m³/s
	13.2 gpm (imperial) = 1 ℓ/s
Mass	1 lb = 0.454 kg
	32.2 lb = 1 slug (US)
Force	1 lb. force = 4.45 N
	(1 Newton = 1 kg × 9.81 m/s²)
Pressure	145 psi = 1 MPa = 1 MN/m² = 1 N/mm²
	14.5 psi = 1 bar
Energy	778 ft lb. = 1 Btu
	1 Btu = 252 calories
	1 calorie = 4.18 Joules
Power	550 ft lb/sec = 1 HP
	1 HP = 0.746 kW
Kinematic Viscosity	1 sq ft/sec = 929 stokes
	= 0.0929 m²/s
Absolute or dynamic viscosity	1 centipoise = 0.001 kg/ms
Temperature	F° = 32 + 1.8C°

Absolute temperature R°
   (Rankine)   = F°(Fahrenheit) + 460
Absolute temperature K°
   (Kelvin)   = C°(Centigrade) + 273
π   = 3.14159
e   = 2.71828

AUTHOR INDEX

SUBJECT INDEX